ALLE ZEIT WACH 1842

Roland Thord (Ed.)

The Future of Transportation and Communication

Visions and Perspectives from Europe, Japan, and the U.S.A.

With Contributions by
Å. E. Andersson, K. Button, H. G. Forsberg, B. Johansson,
P. Kiselev, K. Kobayashi, K. R. Kunzmann, T. D. Larson,
K. Lindahl-Kiessling, H. Sandhäger, R. J. Schonberger, L. Schotte,
K. A. Small, F. Snickars, R. Thord, W. Westerhuis, K. Yoshikawa

With 44 Figures

Springer-Verlag Berlin Heidelberg New York London
Paris Tokyo Hong Kong Barcelona Budapest

Dr. Roland Thord, TEMAPLAN AB, Applied Systems Analysis
for Industry and Government, Sandhamnsgatan 81,
S-115 28 Stockholm, Sweden

Cover illustration by class 6A in Oppeby school, Nyköping, Sweden, the winner of the Drawing Festival 1991 arranged by the Swedish National Road Administration among Swedish school children 12 to 13 years old.

"The picture symbolizes today's longing for sun, freedom and adventure"

(Motivation given by the Jury)

ISBN-13:978-3-642-78033-2 e-ISBN-13:978-3-642-78031-8
DOI: 10.1007/978-3-642-78031-8

Softcover reprint of the hardcover 1st edition 1993

42/7130-543210 – Printed on acid-free paper

"Could you tell me please which way I ought to go from here?"
"It depends a good deal on where you want to get to", said the Cat.

("Alice's Adventures in Wonderland", by Lewis Carroll)

CONTENTS

FOREWORD 1
Per Anders Örtendahl

INTRODUCTION 3
Roland Thord

PART I: VISIONS OF FUTURE ARENAS 9

INFRASTRUCTURE AND THE TRANSFORMATION TO THE C-SOCIETY 11
Åke E. Andersson

THE EMERGING NEW ARENA OF TRANSPORTATION AND COMMUNICATION IN JAPAN 27
Kiyoshi Kobayashi

THE FUTURE OF EUROPEAN TRANSPORT 45
Kenneth Button

TAMING THE WILD ROADWAYS: TEAMWORK, COMMUNICATION, AND SYNCHRONIZATION ACROSS SPACE 85
Richard J. Schonberger

PART II: REGIONAL PERSPECTIVES 95

THE CENTRAL EUROPEAN PERSPECTIVE 97
Heinz Sandhäger

THE EASTERN EUROPEAN PERSPECTIVE 109
Igor P. Kiselev

THE NORTH AMERICAN PERSPECTIVE 119
Thomas D. Larson

THE JAPANESE PERSPECTIVE 137
Kazuhiro Yoshikawa

PART III: NETWORK PERSPECTIVES 173

TRANSPORTATION INFRASTRUCTURE, PRODUCTIVITY AND GROWTH 175
Börje Johansson

A SYSTEMS VIEW OF INFRASTRUCTURE PLANNING 191
Folke Snickars

SOME ASPECTS ON INTERDEPENDENCE OF COUNTRIES AND REGIONS OF IMPROVED TRANSPORT LINKS 201
Willem Westerhuis

TRANSNATIONAL PLANNING EDUCATION AND THE INTERNATIONALIZATION 205
Klaus R. Kunzmann

PART IV: ENVIRONMENT PERSPECTIVES 221

THE ENVIRONMENTAL SITUATION IN EUROPE 223
Lennart Schotte

TRANSPORTATION AND THE ENVIRONMENT 231
Kenneth A. Small

HUMAN ASPECTS ON COMMUNICATION AND THE INFRASTRUCTURE 247
Kerstin Lindahl-Kiessling

TECHNOLOGY FOR IMPROVED TRANSPORTATION AND ENVIRONMENT 257
Hans G. Forsberg

FOREWORD

We all know that networks are fundamental prerequisites for prosperity and production. Transportation and communication are indispensible to society, they are the elements which bind all economic systems together. Without networks and communication all social and economic life will be reduced to isolated phenomena. Therefore, transportation can't be assessed in the same way as other services.

A smoothly functioning system of communications is also a prerequistite for social and economic integration between separate geographical regions. The modernization of the infrastructure is therefore an urgent task and a precondition for carrying out the whole of Europe's ambitious political, economic and social agenda. Since the need for communication and transportation does not know any national borders, the functioning of the networks needs to be adopted to this new economic and political geography.

Congestions of cities, highways, railroads, airways and telecommunications must be tackled, if precious working, commuting and leisure time is not to be wasted and heavy burdens on the environment avoided. European traffic, is for example, expected to double within the next twenty years. In certain transport modes the growth is expected to be even faster – air passenger transport doubled in 10 years and goods transport on roads doubled in 15 years.

Still the level of investments in the European transportation network is less than 1 percent of the G.N.P., or less than half the level needed to meet the development of the demand on the transportation systems. New approaches are needed for deciding where to invest and how networks are to be operated: not through piecemeal connections of existing national networks and clumsy interfaces between different modes, but by setting ambitious functional transport and communication requirements internationally along with an appropriate mix of modes. This in turn not only requires new intermodal links, but also a conceptual revolution which recognizes that the communication needs are in fact served by a single – if still fragmented network – which needs to be looked at in its entirety.

The 150 year anniversary of the Swedish National Road Administration offers a great challenge as well as an opportunity to analyze and structure the future development of transportation and communication, and to highlight their significance for mankind, environment, society and economy.

It is a great honour and pleasure for me to present this book as a report from two seminars arranged by the Swedish National Road Administration in cooperation with the International Road Federation, the Royal Swedish Academy of Engineering Sciences and the Royal Institute of Technology in Stockholm. Nearly 400 participants from 24 countries on all five continents took part in the seminars, representing industry, road users, universities and research institutes as well as governmental bodies responsible for transportation and communication.

As well as thanking each of the above organizations for their support and cooperation, I wish to thank each of the contributing authors for their excellent work both during and after the seminars. I also want to thank the editorial staff at TEMAPLAN AB for their skilful preparation of the complete manuscript and Dr. Roland Thord for editing the book.

I do hope that this book with its contributions given by internationally renowned scientists and administrators will contribute to our common efforts in developing transportation and communication for the future.

Dr. Per Anders Örtendahl
Director General of the Swedish National Road Administration

INTRODUCTION

Roland Thord
Temaplan AB
Sweden

Transportation and communication systems are the elements which intergrate human activities in space and time. They record our past and will be the important determinator of our future.

Transportation and communication are closely related infrastructural elements in our society. They are used to intensify and facilitate exchange in national and international economies. Furthermore they are the most importants prerequisite to increase the level of human interaction and mobility. Mobility and transportation in time and space gives us the possibilities to seek knowledge, experience and social contacts. Mobility is a central and fundamental feature of human values.

Industrialization, urbanization and the mobility created by motorization have changed living standards and the structure of society. Furthermore, a strong internationalization has characterized the development. A renewed phase of prosperity and growth will also demand new transport and communication technologies, as mobility, information and material wellbeing are symbiotic ingredients of modern society.

As long as mankind has had the need to move on land between different places, there has been some sort of road or, initially, path. In the beginning of the century the automobile was born. Paved roads reduced the time-space dimension of modern societies. The automobile age is perceived as one of the preconditions for modern industrial development. In this perspective the car became a synonym for individual freedom and mobility.

In a society characterized of growing flows of increasingly valuable resources the ability to coordinate the flows into new combinations and products will be one of the most strategic things to handle in the future. Transportation and communication in modern times - movements of both people, goods and information - consists of many complex systems. This means that the admission to the flows instead of solid resources or primary products will be more and more important in an economic as well as in an intellectual and social sence.

With this perspective transportation and communication will be crucial for the renewal of production. Free trade and mobility will especially in the long run create new possibilities for production. In this process the conditions for old and established flows and products will be pulled down. The transport system and the communication system will act as the prerequisite for the structural change.

The focus of this book is to analyze and structure the future development of transportation and communication and to highlight their significance for society, economy and for the people and their environment.

In Part I – VISION OF FUTURE ARENAS – four visionary perspectives of future transportation and communication are presented. The four contributors provide us with their views of the development in production as a consequence of changes and rapid developments in the transportation and communication system. First *Åke E. Andersson* gives an overview of the infrastructure and the transformation to the C-society and its arenas. He discusses the slow but steady change of the arena of economic and social life during the twentieth century. He point out the C-regions in Europe and discusses its characteristics – expansion of creative activities, cultural activities, modern communication systems and centers of cognitive activity. This new arena for economic and social life and its relation to transportation and communication is also discussed by *Kiyoshi Kobayashi*. He stresses the responsibility Japan is shouldering in creating a new order for the global network. *Kenneth Button* discusses the most probably developments in European transport over the next decade or more. He argues that only limited technical changes will occur over this period and that the major forces influencing transport developments will come from institutional factors, market influences, changes in social priorities and from increased utilization of existing high-technology. The final paper in part I is written by *Richard J. Schonberger*. He is advocating the manufacturing renaissance and its influence upon the world's road networks. Forces are at work to create partnerships among producers, final sales points, intermediate freight carriers, and regular agencies, including roads bureaus, which extend the manufacturing renaissance to related sectors. Operations will be uniformly simple, quick, capable and at low-cost.

In Part II – REGIONAL PERSPECTIVES – representatives from three continents presents their visions on desirable and attainable transportation and communication systems in the future. This kind of regional presentations is of utmost importance for an efficient use of the world societies resources in developing the systems for transportation and communication. *Heinz Sandhäger* gives his views from a Central European perspective. He discusses the work and topics for the first Traffic Infrastructure Plan for the United Germany. *Igor Kiselev* reviews the situation in the European part of the former USSR – todays Russia. Here it is obvious that the future must consist of hard work and huge investments in order to solve the shortage in transport service and to meet with global standards. *Thomas D. Larson* presents the general framework for specific programs for each major mode of transportation and communication in the U.S. for the next decades. Progress in both areas are helping to draw the world closer together and offers a promise of a better quality of life as we approach the 21st century. *Kazuhiro Yoshikawa's* presentation first of all covers the development of transportation and communication in Japan during the last four centuries. He also discusses the relationship between post-war economic growth and the development of transportation and communication in Japan and finally reviews the future of transportation and communication i Japan.

In Part III – NETWORK PERSPECTIVES – the interaction between different transportation and communication modes is considered as a fundamental process of economic integration. Economic growth in nodes generates growth of demand for transportation and interaction which trigger investment in network infrastructure and transport equipment. *Börje Johansson* analyzes the interdependence between infrastructure capital and other economic resources. He is emphasizing on questions concerning how fast does the economic system react to changed infrastructure conditions and which infrastructure variables affect productivity and growth most profoundly. *Folke Snickars* takes the network perspective as a starting point for a discussion of infrastructural planning. The days when decisions regarding infrastructure were a purely national affair are over. *W. Westerhuis* discusses the interdependence of countries and regions on improved transport links for economic growth. The only way to cope with this situation is to tackle the development of international networks on a supranational level. *Klaus R. Kunzmann* presents the prerequisite for a transnational planning education. The need for this is obvious in a world were policies and standards are formulated at an international and transnational level. Many of the possibilities and disadvantages with this network society can only be developed or solved in cooperation over national and regional borders.

In Part IV – ENVIRONMENT PERSPECTIVES – the contributors highlights the environmental implications of transportation and communication. This dimension has become more and more important and will in the long run be as fundamental a prerequisite for prosperity and production as the transportation networks itself. *Lennart Schotte* argues that the environmental problem is international, the scale, however, of each instance of decline is local to regional. The spectrum of air pollutants differs from region to region. Thus, while strategies to combat undesirable impacts must be related appropriately to the local or regional scales, the strategies must be set firmly in a context of the international arena to take account of transboundary pollution. In his presentation *Kenneth A. Small* assesses the most likely types of future responses to air pollution, greenhouse gases and traffic congestion. His focus is especially whether responses will be technological or behavioral. *Kerstin Lindahl-Kiessling* is advocating the human aspects on communication and the infrastructure. According to her we need a revolution in our attitude to mobility and transportation, using all the means there are to finally achive the environmentally sound solution to every transportation problem. *Hans G. Forsberg* is overviewing new technology and implementation of already known techniques which can give important contributions to a better environment. He distinguishes two trends. One concerned with efforts to improve transport efficiency and reduce fuel consumption and thus the emission of environmentally damaging substances. The second trend is concerned with the combustion process itself and the efforts to render emissions harmless. He finally remarks that environmental and traffic policies are integral parts of an overall discussion about the problems of modern society.

The opportunities for sustainable, long-term environmental development are to be found in active interaction between international, regional and local conditions and solutions. The environmental problems in the most densily populated areas will require a new system approach to traffic. New principles for the coordination of housing and industrial estates with the transport and communication system are available and must be used.

The long-term societal responsibilities and opportunities are related above all to our common active contribution as researchers and as practitioners. However we should not ignore the responsibility we have as individuals and the values that guide us in our daily behavior.

PART I:

VISIONS OF FUTURE ARENAS

INFRASTRUCTURE AND THE TRANSFORMATION TO THE C-SOCIETY

Åke E. Andersson
Managing Director of the Swedish Institute for Futures Studies
Sweden

INTRODUCTION

The Swedish economist and Nobel laureate Gunnar Myrdal was once asked about his definition of economics. His answer was somewhat circular and very laconic: "Economics is what economists do". This implies that economics is a moveable object with changing points of gravity and emphasis. For a very long time, infrastructure was no central part of economics. Judging by the subject indices of standard American textbooks, roads, railroads, and networks are rarely mentioned. Infrastructure is hardly analysed and sometimes not even defined.

In the beginning of the 1980's fashions changed. Increasingly the economists turned to an analysis of the difference in growth rates and change of economic structure between Germany, Japan, Switzerland, and USA, Canada, and the U.K. Obviously, the share of investments in GNP could explain a large part of the growth differences, but more and more frequently investment in roads and other infrastructure apparently had an impact on the pace of economic development. Some of the economists, notably Aschauer, even dared to investigate short term statistical association between infrastructural investments and the rate of general economic growth. One example is given by the diagram on next page.

Similar illustrative diagrams have been published by Folke Snickars of the Swedish Royal Institute of Technology.

Personally, but based on methodological considerations, I do not consider this category of statistics convincing evidence of the relation between infrastructure and economic growth or development. At best, such statistics can indicate the consequences of

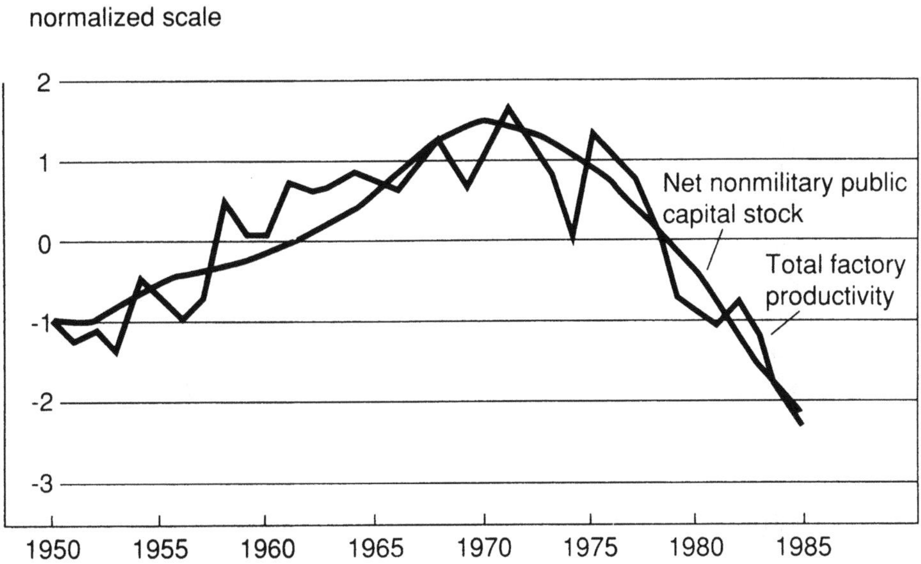

Figure 1. *Changes in total factor productivity of net public capital. (United States). Source: Aschauer (1989).*

severe infrastructural bottlenecks and the responses to investment in infrastructure similar to severe constraints on infrastructure availability. Economic development is a secular issue and infrastructure is only meaningful in a secular context.

DEFINING INFRASTRUCTURE

Infrastructure, be it tangible or non-tangible (material or non-material), is analogous to the foundation of a building. A minimal foundation is characterized by the general importance of all of its parts. Removing one stone would have serious repercussions for the stability of any part of the building erected upon it. Similarly, removing one road link from a minimal road network would have serious consequences for the efficiency and stability of all flows in the network. This implies that not all links of a non-minimal network have the characteristics of infrastructure. A private road is not, to any significant degree, to be looked upon as infrastructure. The first and primary characteristic of infrastructure is publicness. To qualify as infrastructure a road must be a public good. We must, however, refrain from the simplistic view that only people directly using a road are the beneficiaries of that road. A couple of counter examples can be given. Increasing the capacity of one road might decrease travel times on a number of other roads and even of the road network as a whole. Even people who are not themselves users of any roads can benefit of an improvement in productivity due to improved accessibility of firms.

The second characteristic of infrastructure is the relatively slow rate of change. Many commodities have a durability of less than one year. Even cars, trucks, and other machinery, have a durability of a couple of decades or less. Most networks tend to remain in use for centuries. Via Appia Antiqua provides a good example. Constructed in the Fourth Century B.C. it still remains a reasonably efficient part of the Italian road network. And roads are part of the infrastructure in the sence that they are a **public, highly durable capital good.**

The following table gives a tentative subdivision of different economic variables according to their temporal characteristics and degree of publicness.

Table 1. *Different economic variables according to their temporal characteristics and degree of publicness*

	Speed of Change	
Effects	Fast	Slow
Private (Individual)	Market Commodities **Games**	Education
Public (Collective)	Information	Networks **Arena**

The further down to the right a variable is placed the more it acts as an arena, constraining the dynamic movements of the fast game of classical economic variables.

INFRASTRUCTURAL CONDITIONS OF THE INDUSTRIAL SOCIETY

It is clear from the above account that the causalities work from the slow processes of infrastructural construction toward the establishment of compatibel economic and social equilibrium structures of society. However, exactly as in earlier periods of economic development, there has been a gradual destruction of the bases of industrial society, a creative destruction provided by the industrial society itself. Although the heavy causality goes from the arena to the games there is a light causality by investment going from the games to the arena.

In order to understand the industrial society, its development and later destruction of the foundation on which it was built, we have to go back to the early nineteenth century. Transportation and communication networks were then to a dominating extent based on the use of water. The networks consisted of newly established ports, a highly developed sailing technology, and canals and rivers which provided limited contact possibilities with inland resources and markets.

This water oriented transportation network combined with Adam Smith's principles of free enterprise, (economic freedom, property rights and efficient division of labor), turned out to be economically powerful. It provided the profits necessary to generate the enormous capital accumulation in the new accessibility centers of England, Belgium, and New England and later on Scandinavia and other peripheries. It also provided surplus profits, sufficient for exploring and connecting new sites of production and resource extraction. Furthermore, some of the surplus profit was used for daring expansions of new means of transportation – for steam engine driven ships and trains on railroads.

The major period of industrialization, occuring during the nineteenth century, thus became the major period of the railroad and steam ship networks. Industrialism and railroad building went hand in hand reaching progressively further into the two North Atlantic continents. In the first wave it encompassed Britain, Belgium and northern France. In the second wave it progressed into the Ruhr and other industrial centers of Germany, into France and other central parts of the European continent. In a third wave it went further toward the periphery of Scandinavia, eastern and southern Europe. Below, we illustrate the penetration of the railroad system into the North American continent during

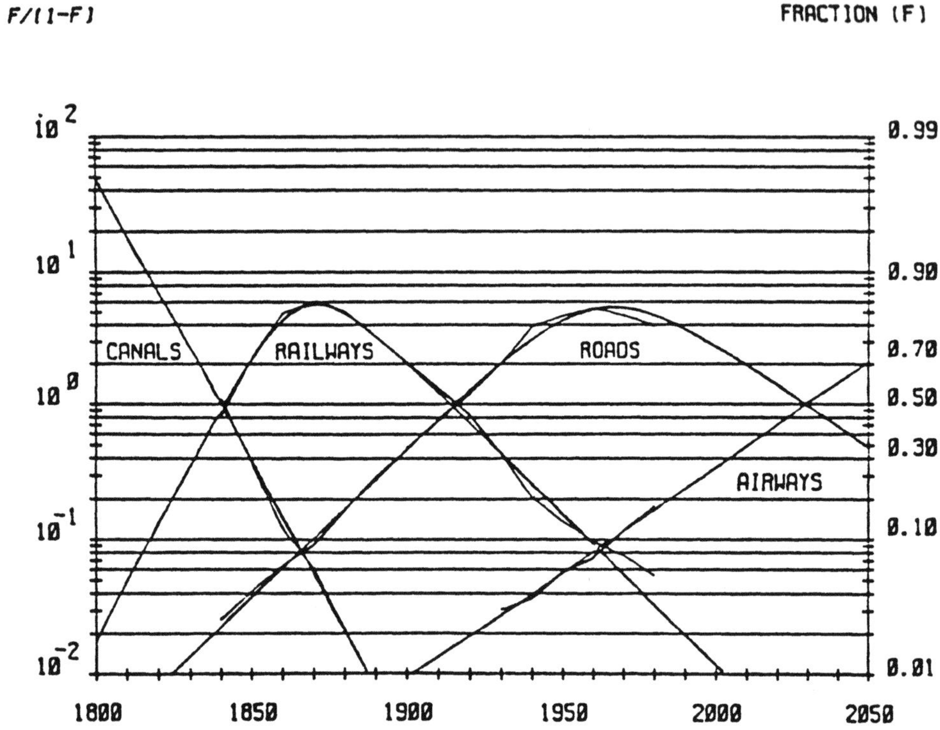

Figure 2. *Relative share of total network lenght of different networks in the USA. Source: Nakicenovic, (1986).*

the nineteenth century.

It took no more than eight decades for the American railroad network to reach relative dominance, counted from the initial building decade. At the end of the 1870's the railroad system provided more than four fifths of the total network lenght of all transport infrastructure of North America. The relative role of the railroads as a network for transportation and communication were similar in western Europe with a couple of decades of delay.

The industrial system being developed before the First World War was based on the following infrastructural conditions:

The education and other knowledge infrastructure grew steadily in Europe and the United States. Compulsary education was practically everywhere introduced before the industrial revolution could occur. The average number of school years of the then advanced countries was approximately three years per capita of the working population in the 1870's, but the variance of education was very large and strengthened the need for a hierarchical organization of society.

Networks for the transportation of goods, people and information had a tree-like structure, which spanned out hierarchically from a few major ports. Some examples of such ports are Liverpool, Manchester, Glasgow, London, Antwerp, New York, Philadelphia and Gothenburg. These railroad/port centers had enormous advantages in terms of accessibility, landrents and profits. Thus, they became powerful attractors of capital, employment and population. They grew into manufacturing metropoles.

Similarly, although on a smaller scale, centers of energy and other mineral resources also became centers of new manufacturing.

Along the Atlantic coast certain regions became natural "points of tangency" between the continental railroad networks and the inter-continental shipping lines. Such continental important locations were primarily concentrated at places such as the river Mersey, Glasgow, London, Hamburg, and Marseille. In all of these points of tangency between the two dominant transportation networks an enormous relative profit potential existed.

Production capital and employment were attracted to these points of transportation tangency. Most of these places grew from being rather anonymous port towns into manufacturing port metropolises. Sea ports which also became manufacturing metropoles of one million inhabitants or more are a typical outcome of the industrial revolution.

INDUSTRIAL ORGANIZATION

The newly established arena of transportation infrastructure required a new organization of industrial and other production activities. The railroad transportation system, ocean shipping, as well as coal fired steam engines all rely on economies of scale and uniformity even in regard to individual loads of transportation. This means that fabrication in production systems based on uniformity, long series, and other aspects of large scale production was favoured by the structure of the logistical network infrastructure. Economies of scale and uniformity in transportation required mass production. The typical industrial company was the factory firm. Thus, Fordism and Taylorism are natural consequences of the sparse transportation networks of the industrial era. Knowledge in the form of education was sparse and skewly distributed and so was information as well. The sparsity and skewness of knowledge and information were the preconditions for the strict industrial hierarchy of decision making and control of production.

The material and non-material infrastructural arena of industrialism forced upon society an industrial organization characterized by:

- Economies of scale in production and transportation;
- Uniformity of products and the strict division of labor down to the individual worker; and
- Hierarchical decision making systems, both strategically and in the daily control of production.

This mode of organization of production and decision making was taken over by cultural producers, public administration units and political parties as well as labor units, by imitation or necessity. Large factory units, central labor units, national opera houses and central government boards are all organizationally mimical consequences of the nineteenth century arena of material and non-material infrastructures.

THE CREATIVE DESTRUCTION OF THE INDUSTRIAL SOCIETY

The industrial society - with its foundation and economies of scale and sparse networks for transportation networks of people and commodities – was a highly profitable system during the development process from the 1850's and onwards. Furthermore, the profits were used for financing further growth by investments. Of course, investments were initially concentrated on machinery, factory buildings and other immediately profitable investment objects.

However, the arena of the early industrial period was obviously not the best for long term development possibilities. Although the railroad system was very advantageous, it was soon realized that it was a too sparse network, often prohibiting profitable use of potential resource and demand pockets. Therefore, some of the investment resources were allocated to the construction of supplementary roads to fill "holes in the railroad niche".

Similarly, the transportation of information by mail was a very slow process, because of its dependence upon trains and ships. A new fast system of air communication was an obvious advantage to industrialists, and by the turn of the century the telegraph and telephone systems began to develop beyond its experimental stage.

Slowly but steadily the organization of the industrial society was treathened by these investments into a qualitatively different system of networks. It was to be a creative destruction but, still, it was destructive. Neither a manufacturing metropolis nor a small manufacturing town can survive the de- and reconstructed infrastructural arena. A fourth logistical revolution is coming. It requires a new locational pattern and a new form of social and industrial organization. It will need new firms and new regions.

THE COMING C-SOCIETY

The arena of economic and social life has been slowly but steadily changing during the twentieth century. Initially, transportation and communication had still to be carried by ships and trains on sparse networks, most suited to large shipments. Energy was also primarily transported by these sparse, scale dominated networks. Economic life was organized so as to use these network constraints as profitably as possible.

Our century has been involved in changes of this arena into a completely new multi-layered set of networks for transportation and communication. As can be seen from figure 2, the long term development of the relative length of networks is towards a high and almost constant share in total infrastructural length for the road network. There has even been a slight downturn in the relative role of the road network in the United States and this is expected to be the dominant tendency over the next two decades. A similar downturn is expected in Western Europe around the year 2000. The infrastructure of relative expansion during the next decades is the same in the US and Western Europe - the airline network. The classical railroad system is slowly declining in relative coverage.

The great question mark is the relative importance of new fast railroad systems. As it is, the current fast train systems do not have the necessary characteristics to seriously impede the growth of the airline network. The fast rail systems will rather be complements to the airline system on such metropolitan combinations, where air traffic congestion does not permit large scale inter-city air transportation expansion. With the technological characteristics, the European fast train systems will play a role similar to the one played by the Shinkansen line of the major metropolitan corridor of Japan.

The current transportation and communication situation is not primarily one of substitution between transportation modes but rather a situation of increasing complexity and complementarity between the different networks. If viewed from above, the development of networks during the last century has been a process of super-positioning of new networks on top of the old ones. The railroads were initially complements to the canals and rivers. After that initial period they became a new super-imposed grid of higher density than the canal system. Later on the road system became a complement to the railroad system. After the initial stage it became an independent network super-imposed upon the canal and railroad networks, becoming increasingly denser than the canal and railroad networks of earlier times. The airline network was initially created to care for transportation between regions with very little economically viable personal interaction. Later on the density of this network also increased substantially and it has become another layer super-imposed on top of the older networks. The telecommunications network can be seen as yet another, particularly specialized, very dense network in the total grid.

One of the most important characteristics of the C-society in contrast to the I-society is the level of mobility. On a per capita basis countries like Sweden and France have increased the daily mobility by a hundredfold. By the end of this century, trends indicate a daily per capita transportation in European countries of 39 kilometer.

The systems of the future will to an increasing extent exploit profits of synergies between these different networks. Air freight systems, trucks, air carriers and telecommunications are necessary ingredients of the whole transport solution. Similarly, in the air transportation of people, rail, road, air transportation and telecommunications are always combined in the profit maximizing and thus surviving logistical solutions.

Table 2. Passenger transport by ground transportation 1988 in the USA and a subset of European countries and a forecast for 2000

Land	Km per day (per capita) 1988	Km per day (per capita) 2000
USA	51	55
Switzerland	40	47
Denmark	33	39
France	32	42
Sweden	32	39
Finland	31	39
Norway	31	39
Germany	29	39
Netherlands	29	36
Italy	28	36
Austria	27	35
U.K.	26	32
Belgium	24	32
Greece	6	10
Turkey	5	10
Western Europe	29	39

Source: OECD Data bases

SMALLER PLANTS WITHIN LARGER CORPORATIONS

With the old sea and railroad networks, the number of infrastructural locations with a large profit potential was very limited. At these mainly coastal high profit points of joint accessibility to intercontinental as well as intra-continental transportation and communication, there were set up large scale, hierarchically organized production units. These units employed a division of labor and economies of scale to the utmost, best exemplified by Henry Ford's American automobile plants. The value of division of labor was further reinforced by the use of Taylor's productivity-enhancing methods.

Mass production of uniform products, the division of labor,

hierarchical control systems, and political and other social mass movements can all be seen as the profit and growth maximizing factors of the industrial system. These organizational factors were determined by the then available networks for communication and transportation in conjunction with a scarcity of information and knowledge. With such an infrastructure, no other profit-oriented structure of production and decision-making could evolve and survive.

The slow and steady evolution of a new multi-layered network infrastructure will give rise to a completely different production, distribution and control structure if efficiency and profits are aspired to. The old infrastructure was sparse and oriented towards large scale activities. The new networks are extremely dense and often require a rather limited size of shipments. In most countries, the main road network is more than twenty times as dense as the railroad network ever was. Telecommunication networks are even denser. Practically any location is thus accessible, at least in Western Europe, large parts of the United States and the developed portion of Asia.

This implies that every locational point with a resource pool is of interest to any corporation prepared to develop production technologies, and dependent upon the limitations on the size of the resource pool of the given locations. Likewise, commodity distribution is organized so as to exploit the demand pool of any location. Organizing the flows between such points is much more important than finding out new clever ways of production, transformation and other industrial processes within each location. In a sense logistics is a more important consideration than production processing.

While plants are becoming smaller and smaller, irrespective of their concentration on production or distribution activities, corporations are getting larger. The evolution of ever larger corporations with ever smaller units of production is one of the important structural changes of our economies. Within these corporations a profit-increasing handling of flows of information, knowledge, products and services by people is becoming the primary planning problem. Two excellent examples of the tendencies of the importance of restructuring production and logistics are the Digital Equipment Corporation and IKEA. These corporations are two typical examples of the emergence of a new economic structure.

It is clear that denser networks for transportation and communication permit smaller scale production and distribution units. Why is there then a simultaneous tendency towards larger corporations? This cannot be understood without reference to the inherent economies of scale in the creation, innovation and diffusion of new technologies and other forms of original knowledge. Technologies and other forms of knowledge are different from other inputs. The use of energy by one firm effectively means the non-use of that energy by another. This is not so with knowledge. If a new technology of production becomes available to one firm, it does not preclude the use of that technology by another or even many other firms. Finding systems of cost-sharing for technologies to be used by many plants is thus an important economizing device in a market economy. The same consideration is true for furniture and other product design. Firms like IKEA and Benetton can spread the costs of new designs across many plants, distribution units and ultimately consumers of their products, if they are successful in pooling a large number of production and distribution units (and customers) within their corporations. Often these firms further reinforce the advantage of a large corporate structure of small units by their heavy use of communication systems, developed for the corporation. The administrative procedures of bookkeeping, inventory control, personnel management and financial transactions also provide important vehicles for increasing the size of the corporation.

What has been said does not preclude the importance of small scale creative high-technology firms. The computer industry, the pharmaceutical industry and other knowledge intensive sectors have experienced a fast growth of small firms along route 128 in the Boston region, in Southern California and in the Cambridge region. One can even argue that much of current industrial dynamics is shaped by such small entrepreneurially active, knowledge-intensive units. But in the long run this does not constitute a final solution. Rather it is a transitory, although ever-valid, phenomenon of economic development. Most of these firms merge with, or in other forms become permanent members of, large corporations and other extensive networks of production and distribution units. Some of them have even been created around a new technological idea, being capitalized in the merger into a large corporation.

The new tendency of the C-society is thus reaping economies of scope by an expanding network of production and sales units, while at the same time reaping economies of scale in the research and development activities, world wide marketing, and soft- and group-ware for corporate networking.

C-REGIONS

An equally important consequence of the emerging C-society is the evolution of C-regions. In the classical industrial society every kind of industrial activity, including engineering research and development, was necessarily located to the coastal regions of maximum accessibility. This tendency is now reversed.

No longer is there a need for coastal location. Synergetic advantages of different networks can be reaped in much larger, often continentally located locations.

In Europe, the last two decades of growth has been scattered in the interior parts of Western Europe, as illustrated by the following map.

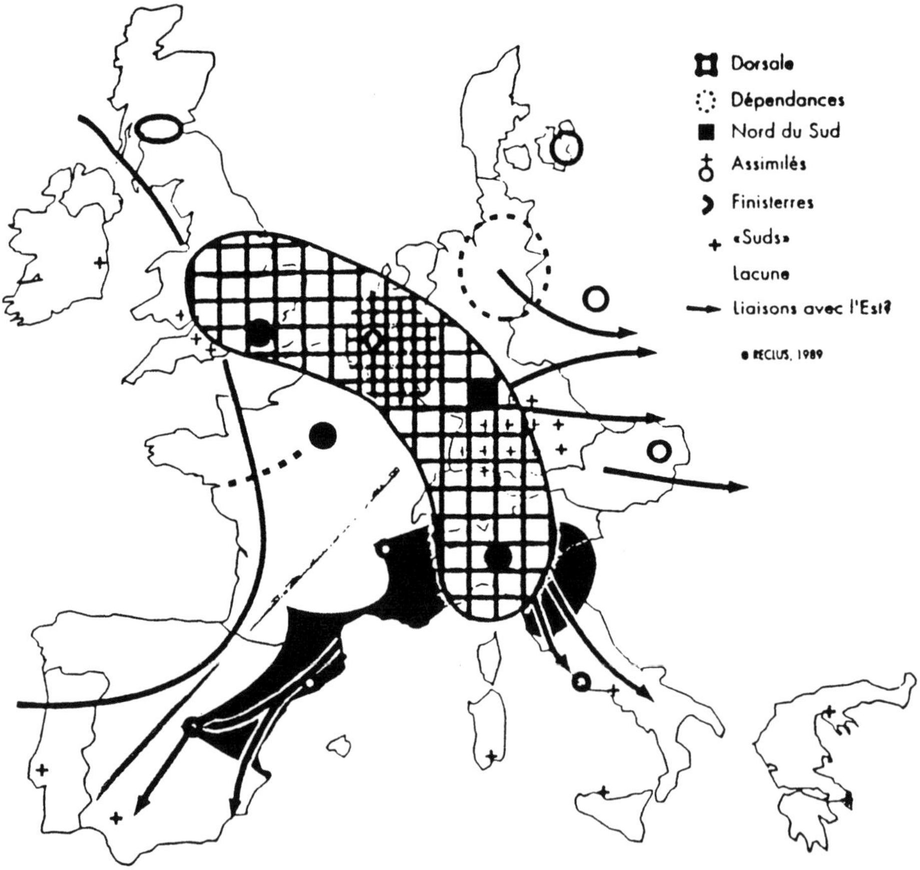

Figure 3. *Urban networks in Europe. Source: Datar "Les Villes Européennes" (1989).*

This growth pattern coincides with the location of modern C-regions. In these European C-regions there is a coordination of expansion of creative activities in research and development, and cultural activities, modern communication systems, on the base of some center of cognitive activity. In most cases these centers of cognitive activities are dominated by some university or some institutes of technology, life sciences and other centers of applied science and higher education.

The following table contains the most important C-regions of Europe with indicators of their creative capacity, as measured by their scientific publication rate, and their communication capacity (as measured by air passenger capacity).

Table 3. *The ten most important C-regions of Europe (Index values)*

	Scientific Creativity	Air Transportation Capacity
1. London-Cambridge-Oxford	100	100
2. Paris	70	69
3. Amsterdam-Utrecht-Leiden	33	25
4. Bonn-Düsseldorf-Cologne	33	21
5. Stockholm-Uppsala	26	23
6. Brussels-Louvain-Gent	24	12
7. Frankfurt-Mainz-Giessen	22	42
8. Munich	22	17
9. Heidelberg-Karlsruhe	22	.
10. Copenhagen	19	21

Sources: Science and Social Science Index, INFORSK, Umeå, Sweden, Air transport statistics 1988.

From this table has been excluded regions which have a large cognitive and communication capacity, but combined with a predominantly old manufacturing industry structure. Examples of such regions are the Manchester-Liverpool region, and the Hamburg region.

The most spectacular change in development pattern is the shift from the English North-West to the South-East with its corridor between Cambridge, London, Oxford and Reading. Other examples of a shifting development are the loss of development potential in the northern corridor from Hamburg to Bremen in Germany in favour of the development triangel Stuttgart-Munich-Zurich. Similar development corridors can be identified in the USA and in Japan.

The most interesting global tendency today is the growing interaction of these C-regions. This is primarily indicated by the growth of scientific and other research corporations during the latest decade.

For many of these C-regions it is now more important to belong to the right network of C-regions than to be an important part of the national institutional framework. The number one priority of these regions is however to develop itself into a dynamic combination of communication, creative and cultural systems.

VALUES OF THE C-SOCIETY

An economic and social structure adapted to the C-arena cannot become viable and stable, unless fundamental values of the population are adapted to the arena as well.

A pleasant society requires, on the on hand, a high degree of individual adaptivity and creativity to be efficient. The need for defence of the natural resources requires on the other hand a hierarchical system of decisions on these highly public matters.

The industrial society could not accept the same degree of individualism in the production system. Rather the strict division of labor demanden a class structure, dividing people into large classes of laborers, controllers and owners. Cohesive of these classes require a class based solidarity principle. Control required preservation and development of hierarchies. Recent studies by Ingelhardt (1990) indicate a changing value structure more compatible with the C-society. This value structure, Post-Materialism, would mean a renewed stress on individual adaptivity, creativity and mobility while it would imply a definite break with the hierarchical principles of decision and control in favour of non-permanence in self-organizing structures continuously evolving so as to solve the perpetually emerging and increasingly diverse public issues.

REFERENCES

Aschauer, David Alan (1989) "Is Public Expenditure Productive?" Journal of Monetary Economics 23, pp, 177-200.

Ingelhart, Ronald (1990). "The Rise of New Political Style", in Urban Challenges. Report to Commission on Metropolitan Problems. SOU 1990:33, pp. 209-235. Stockholm. Allmänna Förlaget.

Nakicenovic, Nebojsa (1989). "Expanding Territories: Transport Systems Past and Future", in Transportation for the Future, pp. 43-66. Springer Verlag.

Smith, Adam (1776). "An Inquiry into the Nature and Causes of the Wealth of Nations." University of Chicago Press, 1976.

THE EMERGING NEW ARENA OF TRANSPORTATION AND COMMUNICATION IN JAPAN

Kiyoshi Kobayashi
Tottori University
Japan

INTRODUCTION

In his stimulating book "Infinite in All Directions", Freeman J. Dyson (1988) investigated the possibilities of new frontiers for "Science of Diversity", and presented a disputable hypothesis on the evolution of organisms. In his arguments, the most fundamental mechanism to catalyzing the evolutionary processes should be sought in the fact that biological synthesis is ingeniously accomplished by interactions between the hardware and software of genes. The modern synthesis reconciles replication with error tolerance by establishing the division between the hardware and software. The hardware must be rigidly designed and error-intolerant in order to maintain the accuracy of replication of genes and to sustain the species. The burden of error tolerance has merely been transferred to the software. The separation between the hardware and software of the organisms enables the species to create immensely complicated systems and to retain the compatibility between the requests for the perfection in its replication processes and those for creating infinitely diversified individuality.

The same type of interactions can be commonly found in the modern society. With the infrastructure of hardware firmly in place and subject to a strict regime of quality control, the software is free to wander, to make mistakes and occasionally to be creative. Junk culture is the rubbish of civilization. Tolerance of junk is one of life's most essential characteristics. In every sphere of life, whether cultural, economic, ecological, the systems which survive best are those which are not too fine-tuned to carry a large load of junk. The primal characteristics of society are to be homeostasis rather than replication, diversity rather than the uniformity, flexibility rather than tyranny, and the error tolerance of the whole rather than the precision of the parts.

Historically, humans have created inumerous break-through in the hardware technology and infrastructure. Language and money are very good examples. These factors of the hardware can function perfectly within each nation. Networks, technology and knowledge are also elements of the fundamental hardware which facilitate any kind of activities in our societies. As Åke E. Andersson stated, sometimes our world evolves with the advent of completely new functions and structures of our society and economy. These drastic forms of evolution which occur when humans create new break-through hardware technology and succeed in introducing new technology into our society are known as logistical revolutions. From the observation of the past logistical revolutions, we know that there are two fundamental preconditions common to all of them. First, the newly emerged technology was fully accepted by societies of the time. Knowledge about new technology became common to all people. Second, more significantly, the advent of new technology led to the increases of the socioeconomic dimensions underlying human activities, which enriched the flexibility and diversity of human activities.

It is very probable that we have been in the midst of fourth logistical revolution, associated with the growth of transportation and communication capacity as well as the growth of knowledge base. This development goes hand in hand with a successive improvement of information technology (Andersson, 1986). The emerging knowledge infrastructural arena can be analyzed by extrapolating Batten, Kobayashi and Andersson's (1988) theoretical work from the firm to the economy as a whole. The new infrastructural arena can be characterized by well-coordinated multi-layered knowledge networks. All flows within an economy can be represented as occurring within a knowledge network comprising nodes such as towns, cities, or metropolitan regions and links corresponding to different transportation and communication modes. The different layers of knowledge networks can be recognized: (1) that of slow-speed transportation networks designed for the movement of resources and goods from manufacturing nodes to consumers; (2) that of high-speed

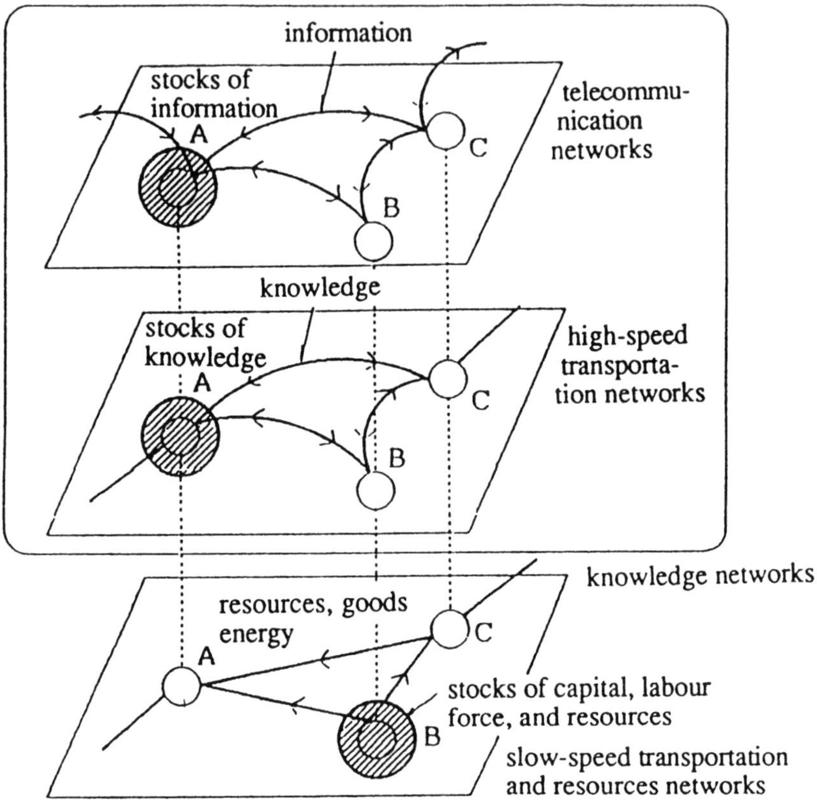

Figure 1. *Multi-layered knowledge networks.*
Source: Batten, et al. 1988.

transportation networks geared to individuals so that synergetic face-to-face business contacts involving negotiation can be facilitated and new knowledge generated by universities and research centers can be transmitted; and (3) that of telecommunication networks intended for the transfer of uncomplicated and routinized information. Technological advances have minimized the friction of distances on flows of goods and messages but it remains an important constraint on the movement of people.

Transportation contacts can be replaced by telecommunications for standardized information but there is so alternative to face-to-face contacts when knowledge is transmitted. Commodity production can be decentralized because of a relative reduction in transport costs, the substitution of labor by robots and computerized production control. Where R&D facilities dominate and face-to-face contacts is essential for the transfer of knowledge, activities are centralized in metropolitan areas. These concepts enunciated by Batten, Kobayashi, and Andersson (1988) underlie the analysis of emerging new arena of transportation and communications in Japan.

PARADIGM CRISIS AT THE TURN OF INFRASTRUCTURAL ARENA

Unpredictability - Is it an irreconcilable enemy for planning?

Since the early 1980s, we have experienced many events which may symbolically indicate the emergence of a new logistical revolution. The first and most obvious consequences of this new tidal wave would be global integration of industrial complexes with fast, spatially and temporally, well-coordinated, but volatile, flows of commodities, people, and information (Andersson, 1986). The basic features of this new wave can be characterized by its nonlinear dynamics, typically described by the concepts of chaos, instability, and volatility of systems. The classic theory of systems dynamics can not provide us well-organized explanation of these phenomena. It has been well recognized that we are in the crisis of systems and planning paradigm. This crisis is related to the basic inquiries on philosophical foundations of systems and planning science and bring about paradigm shifts in the broader field of natural and social science. It has become evident that recent advances in dynamic theories have caused some doubts about the possibility of "understanding", "planning" and "controlling" of social systems in their classical sense of the terms. The new theories provide us a deeper understanding of the changes of human behaviours and social, economic and institutional systems. They taught us that it would be very unlikely to plan and construct a resilient or robust society that could fully accommodate the unpredictable swings of logistical systems. Quite naturally, here arise the following inquiries: "Should unpredictability always imply the uncontrollability of socio-economic systems?" "Should we have be irrational in planning and policy making if there is no way to control the systems?"

East-Asian Understanding on Unpredictability

For Japanese as well as other East-Asian people, the concept of "unpredictability" has been integrated as part of daily ways of systems thinking in a quite natural manner. Unpredictability has been the basic theme of the philosophy on systems thinking in East-Asian tradtion. The earliest document aiming at a systems approach to decision-making in the world was written in the second millennium B.C. (Churchman 1979). This very early attempt was the I Ching, and it was also translated into English with the title of "Book of Changes". The I Ching consists of commentaries on the original writings, so that there are several authors of the book. The I Ching makes some fundamental judgments about the nature of reality. Essentially it assumes that the reality of decision making can be compartmentalized into sixty-four basic possibilities, each of which itself is rather general in nature. In each of these possibilities certain attitudes and guidelines are appropriate for the decision maker. For each possibility, the I Ching develops a model which provides a way of determining how a situation will change.

The I Ching's idea of changes can cope with a really remarkably modern concept of dynamic theories. The I Ching tells us that changes sometimes can be horrendous, judgment is to wait and people should be perseverant. The I Ching's method fails to provide any scientific directions for action, but it can offer general precautions, warnings, exhortations and the like, which the most modern decision-maker might apply to his own framework-making. In this regard, the I Ching has long been serving as a more helpful guide than prescription written in a precise and unambiguous language. Qualitative loose guidelines have proved to be more helpful for decision-making than quantitative hard prescriptions. The "rationality" underlying the persuasion of the I Ching is a basic clue to understanding the reasons why the I Ching's models of decision-making have been attracting human minds over two thousands years and more. The rationality is also the basic precondition of what we now call scientific theory. In obvious, for example, that economic theory was generated based on the concept of a "rational man". The theory is valid because of the way in which it defined rationality and not because it describes the realities of human being.

Controlling the Chaotic World

Major disciplines of planning science and engineering have been more or less influenced by the tradition of "Symmetry These" elaborated by the school of logical positivism. Economics has also been under the influences of Alfred Marshall's understanding of this thesis. In this tradition, "prediction" should be the synonym of scientific explanation. It is considered that a powerful theory which can explain the past changes of a social systems should be capable of predicting its future changes.

The modern dynamic theories have brought doubts to this scientific paradigm. They showed us that a very simple model has power enough to reproduce one of the most complex and chaotic behavior of systems. The link between "prediction" and "explanation" is the very theme of my inquiry into the planning paradigm crisis. Two alternative options to cope with the crisis are available. One is to maintain the classical way of planning thinking by sticking the principles of "Symmetry These". The other is to recognize that "unpredictability" is an essential aspect, and to begin to invent a methodology of which complexity, nonlinearity, instability and unpredictability is a central aspect. Here may arise a question: "Will this methodology be scientific?" If you stick to the assumption that "classical observation" is the basis of science, the answer is negative. If science means the creation of relevant knowledge about the human being and society, the answer is affirmative. The methodology of the second option is essentially based on a schema that can be modified as scientists learn much more about the reality of the behavior of human being and society underlying the currently emerging evolutionary change of social systems as the I Ching tried to do so several thousands years ago. It simply requires a deep knowledge of the nature of dynamic aspects of human's creativity, interactions and communications.

The unpredictability of social systems does not necessarily means to lose the controllability of technology and social systems. To illustrate the controllability of chaotic systems, let us imagine a billiard game. Balls on a billiard table show very chaotic movements. The slight change in angles to hit a ball may result in the completely different movements of the whole balls. It seems to be impossible to predict precisely the movement of each ball.

However, even in the chaotic world like a billiard game, there is one sure thing that we can precisely predict: we can surely anticipate that all balls will be on the billiard table as far as one does not hit a ball too roughly. The billiard table ensures the viability of billiard games. The table in billiard games may correspond to what we call "infrastructure" in our chaotic world. This simple illustration may well explain what should be equipped with to control the unpredictable phenomena.

One of the plausible and promising way to govern chaotic phenomena is to create counterpart systems (infrastructure) which can absorb or integrate the unpredictable fluctuation of the original system. The creation of modern enterprises was the greatest innovation to overcome the instability of the capitalist economy. The security markets are also significant devices to absorb the price volatility in stock resources. Technological and institutional innovation are fundamental solution to cope with the instability of the systems. New infrastructural planning policies in our chaotic world should be sought through efforts in developing the qualitative innovation of systems, institutions, and societies.

NEW INFRASTRUCTURAL ARENA IN JAPAN

A Brief Retrospect on Regional Policy

The Japanese trajectory of the development of transportation and communications are fully described in Yoshikawa (1991). It has always gone hand in hand with economic development in Japan, and was simultaneously the history of industrialization. In this section, I just try to give a simple retrospection on the development of general regional planning policies. Japan has been subject for comprehensive national development plans. The government has formulated four Comprehensive National Development Plans since the first plan was formulated in 1962. In the latter half of the 1950's, Japan entered a period of high economic growth. During the high economic growth era, a large number of young people migrated to Tokyo and Osaka, causing serious problems such as over-concentration and regional disparities. The first Plan aimed at preventing excessive growth of metropolitan regions and correcting the regional disparities, the plan defined those rural areas where it was difficult for residents to make use of the existing accumulation of public and private capitals and advocated establishment of industrial development bases in such areas.

The rapid rate of economic growth brought about a further concentration of population in metropolitan areas. In order to cope with these problems, the New Comprehensive National Development Plan formulated in 1969, aimed at more balanced use of the national land by extending over the entire nation development possibilities. It set forth the concept of large-scale development projects including the new national networks of transportation and communications which would consist of the national axis linking seven major cities from Sapporo through Tokyo to Fukuoka and local networks branching from axis. This idea has been nearly realized.

Entering the 1970's the Japanese economy shifted to a path of stable growth. The Third Comprehensive National Development Plan was formulated in 1977. This plan strongly recognize that regional development has let behind an integrated arrangement of local living conditions and adopted the Integrated Local Settlement Policy. This policy proved effective during 1970's as the influx of population to the three metropolitan areas subsided, which asserted the progress of local settlement. Each region showed a growing tendency to local development based on its own initiatives as well as actual improvement of living conditions.

Emergence of New Infrastructural Arena

Entering the 1980's, however, the population renewed its inflow to the Tokyo metropolitan area, coupled with the uni-polar concentration of advanced city function there. Meanwhile, many rural areas have witnessed aggravated unemployment as a rapid restructuring of industry led to a stagnation of industries with matured technologies and producing export-oriented goods. Rural population is again on a down trend (See Figure 2). Many factors have worked together behind the emergence of uni-polar type of national land structure. Among others, the integration of nation wide human, knowledge and materials networks into the part of global networks would be the major driving force. Tokyo is playing the unique role within the nation as an international focal point.

Looking toward the turn to the 21st century, some of substantial socio-economic changes are foreseen, which symbolically show the emergence of new infrastructural arena: the progress of technologi-

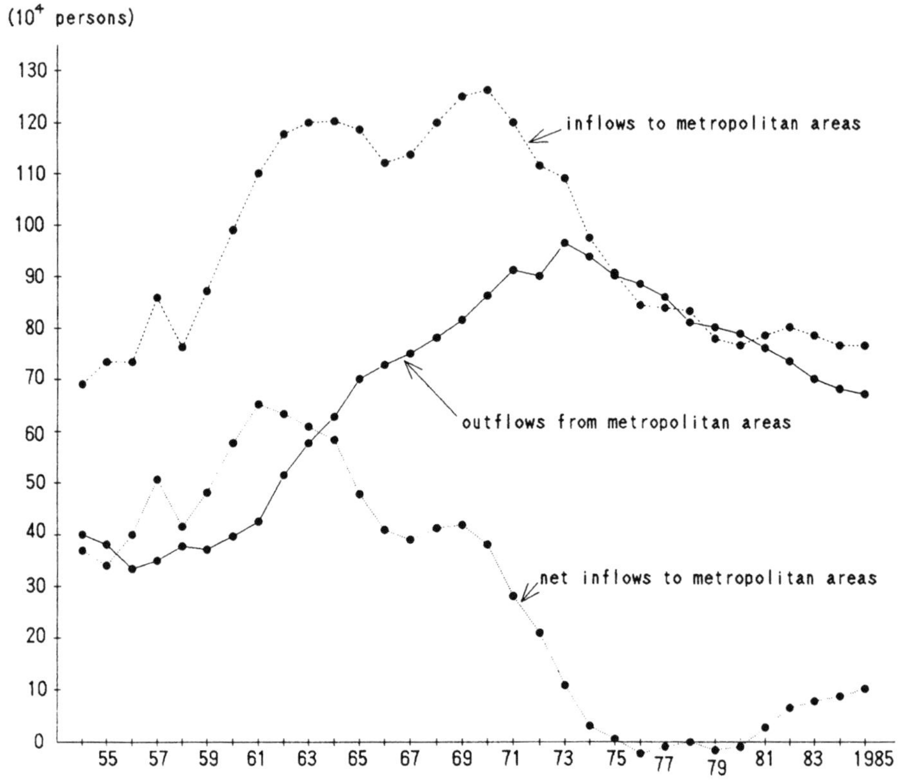

Figure 2. *Migration between metropolitan and non-metropolitan areas.*

cal innovation and a trend toward a knowledge society would bring about new possibilities for the economy and society, together with the rapid metabolism of industrial structure; new life styles operated on the weekly bases have emerged; need for higher quality of life and safety for citizens would become intensified under the trend of aging and relatively stable growth of the population; and Japan would be thoroughly integrated into an international society.

Since the late 1970's a denser, multilayered communications and transportation networks have been progressively introduced accompanying the emergence of knowledge society, progress in global integration of networks, the advent of an information network society, an increase in free time, and in area, and "multi-habitation". People have begun to seek various relations with others, which has resulted in the rapid increases of the human mobility (See Figure 3). Thus Japan has entered an era of active interactions. Arrangement of principal transportation systems as well as information and communications systems is requested to obtain more flexibility, reliability and diversity in routing for the movement of people, goods and information. It is recognized that connections between major cities of the country including province and prefecture center cities should be strengthened through nationwide extension of high-speed transportation systems. This will enable people to travel round-trip between major cities within one day (nationwide One-day Traffic ranges). To cope with the change of circumstance, it is recognized that the major initiatives for regional development should be shifted from the central government to area's own hands.

Transportation and Communications in New Arena

The new infrastructure (a structurally-complex logistical system) is permitting diverse and volatile flows of commodities, information, persons, and knowledge and global industrial complexes to capture economies of scale, scope, and networks by creating synergetic networks, marketing and product strategies in response to their R & D activities. The highly educated students with more than 16 years schooling would be the major preconditions to realize these types of economies.

The rapid increases in demands for high-speed transportation with relative declines for slow-speed transportation can be observed in various fields. In physical exchanges, international cargo is expected to increase to 810 million tones in 2000, 1.2 times as mush as that of 1985, reflecting a favourable increase in world trade although exports and imports will become high value added and light resulting from sophistication of industrial structure and international division of labor. Slow-speed transport networks are focused on the major three marine foreign trade hubs – Tokyo Bay, Ise Bay, and Osaka Bay. The existing slow-speed networks set the logistical standard for Pacific Asia with its just-in-time distribution systems. Transnational corporations, caught in the economic downswing, have revitalized their physical distribution management systems to take full advantage of international container movements. Due to the increasing scarcity of laborforce associated with goods movements, institutional and technological innovation should be catalyzed to increase the efficiency of physical exchanges (National Land Agency 1987).

Freedom of interactions will expand thanks to improvements in high-speed transportation systems enabling varied activities throughout the country. One-day accessibility population represents the possibility for a person living in an area by a one-day round-trip. The national average of one-day accessibility population was about 35 million in 1970 and increased to about 47 million in 1985 to show substantial increases in the possibility of interactions in 15 years. This number will reach about 59 million or nearly half the total population of Japan in the future. In order to balance the demands and supply of passager movements the trunk transportation systems, i.e., a total of 14,000 kilometers long of expressways and rapid passenger transport systems using linear motor technology (the Chuo Linear Expressway bridging the distance between Tokyo and Osaka in one hour) will be in existence.

Human exchange with the rest of the world will increase steadily. The number of foreign people who visit Japan in 2000 will rise to 6 to 8 million (2.3 million in 1985). Japanese travellers to overseas will increase 10 to 12 million (4.9 million 1985). Foreign residents will increase to more than 600,000 in 2000 on average annually (National Land Agency, 1987). The short-term means to cope with the increasing demands for human movement is: (1) to expand the existing international terminals at Osaka, Haneda, and Narita; (2) to redevelop local airports; and (3) to build the New Kansai International Airport and Third International Tokyo Airport open 24 hours per day. A multilayered communications and transport infrastructure is expected to sustain the increased amount of flows in the long run. A new set of airports will be constructed based on a set of superhubs of air transport corridors. Feeder links will be provided by: (1) super trains (maglev linear motor systems) and (2) tilt-wing or tilt-rotor aircraft capable of the vertical taking-off. Through facilitating point-to-point contacts between dispersed locations, telecommunications and air transportation will have intensified the hub and spoke character of the global networks.

Information exchange will expand significantly against the background of progress in the international integration of business activities and personal living as well as progress in communications networks. As a result, the volume of international information exchange will grow in 2000 to a size 20 times as large as that in 1984 (National Land Agency, 1987). The advent of new information technologies like satellite, optical fiber communications, artificial intelligence and a radio frequency spectrum

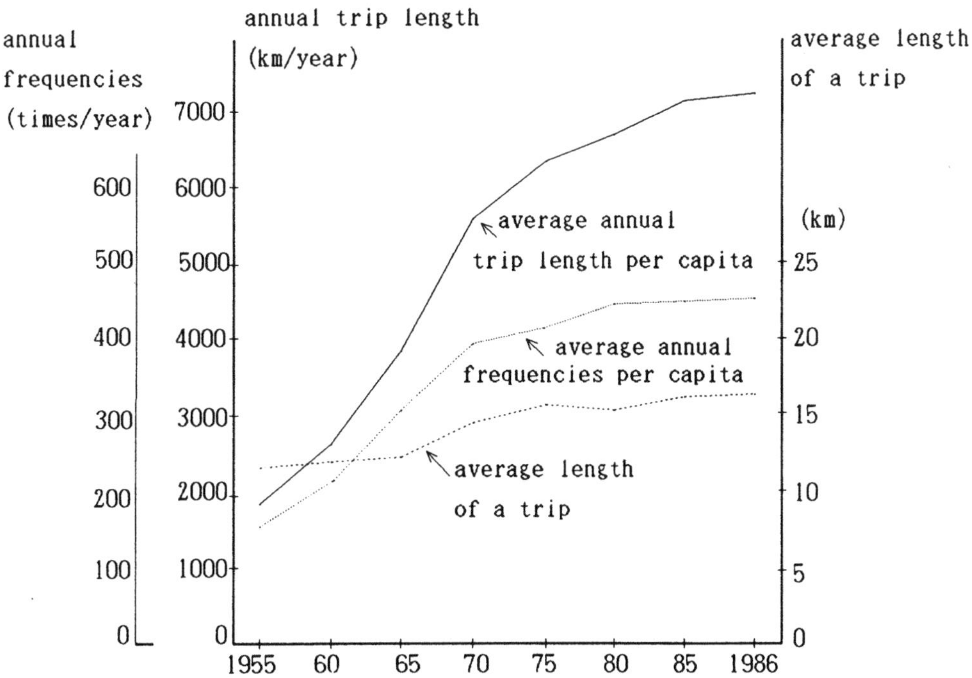

Figure 3. *The increase of mobility in Japan.*

together with the higher education on information science enables to overcome the geographical distances and to provide enormous capacities for interactions. While the distance lays a role of decreasing importance in the exchange process of data and information on the telecommunication networks, the distance friction on the transportation networks now and in the future will be considerable in the movement of people. As shown in Figure 4, telecommunications and transportation networks exhibit the different topological features in terms of accessibility measures. (Accessibility measures represent the distribution of knowledge availability in a simple way that imposes a very clear structure upon the relationship between activities and their environment.) Japan shows the bi-cores structure on the face-to-face oriented accessibility, while she does the uni-polar structure in terms of

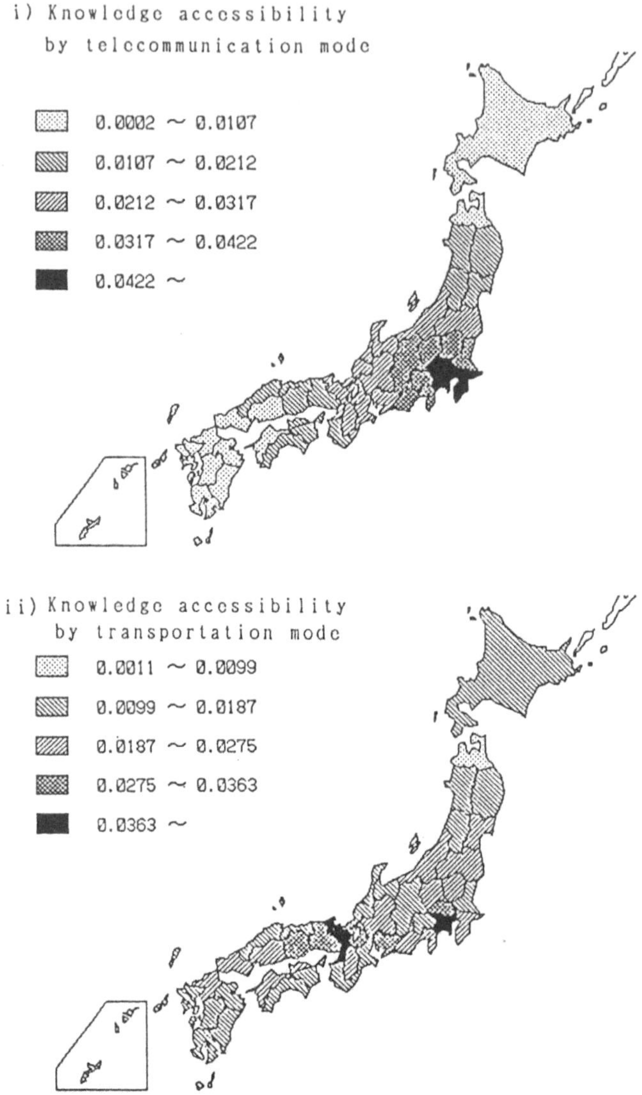

Figure 4. *The spatial topology of accessibility structure (in 1987).*

Note: Accessibility measures both by telecommunication and transportation modes are given by:

$AC_{telecom} = \Sigma_j \, exp(-0.0037 d_{ij1}) \; G_j^{2.9021}$ and

$AC_{transportation} = \Sigma_j \, exp(-0.0239 d_{ij2}) \; G_j^{0.1504}$, where d_{ij1} is the interregional telephone cost between region i and j (yen), d_{ij2} is the interregional travel time (minutes), and G_j is the population in region j. Accessibility measures are estimated based on the data from the interregional information flow survey (National Land Agency 1987).

telecommunications accessibility. These basic characters of the multilayered knowledge networks are now under transformation processes due to the global, especially pan-Pacific, network integration.

PAN-PACIFIC INTEGRATION

The global knowledge networks can be characterized by the discrete features of their structure. Through facilitating point-to-point interactions between spatially dispersed locations, telecommunications and air transportation networks have been integrated as global logistical networks. The global transportation and communications networks in the Pacific Rim have also the discrete structure and will have the hub and spoke character. Figure 5 describes the major non-stop flight routes in the Pacific Rim in August 1989 (Rimmer, 1991). There can be found strong connections between Japan and Pacific Asian countries. There was main corridors between Jakarta and Taipei, in which Hong Kong played an important pivotal roles stemmed from connections with China, and Bangkok's prominence arose from its function as a fueling base on the one-stop Australia-Europe flights.

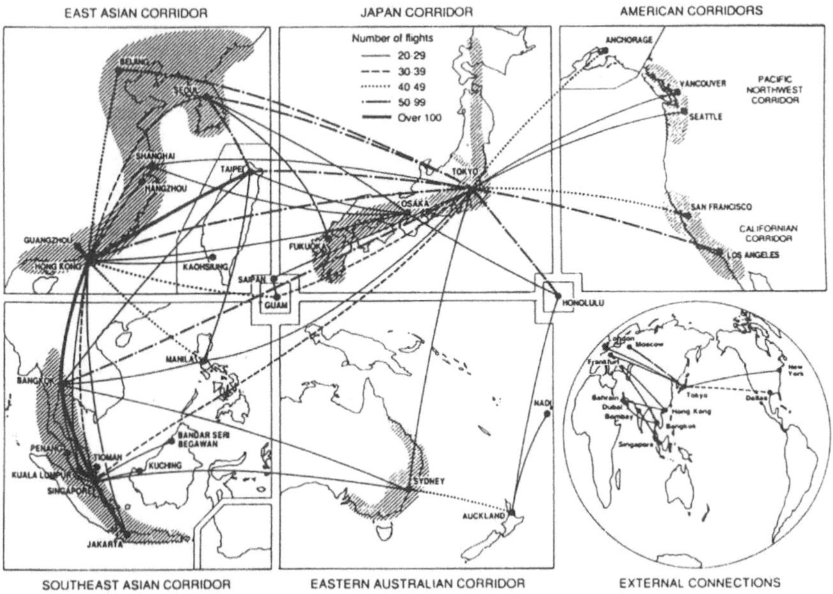

Figure 5. *Twenty or more non-stop flights in the Pacific Rim. Source: Rimmer 1991.*

The most impressive feature, however, was Tokyo's dominating role as the principal gateway to Pacific countries resulting from its strong local markets within Japan and its being a fuel stop on trans-Pacific flights. Within these corridors network hubs are benefiting from being national and international finance centers and headquarters of global network corporations (Rimmer, 1991). They are trying to capture potential synergies and collective knowledge inherent in the dominant communications and transportation networks.

An examination of fiber-optic cables highlights investment in telecommunications infrastructure. These submarine fiber-optic cable networks are being laid to connect major cities in Pacific Rim. Following the explosion in demand for international telephone calls the new network is designed to overcome a feared grids in satellite communication and microwaves. As shown in Figure 6 the entire network will interconnect major cities in the Pacific Rim (Rimmer 1991).

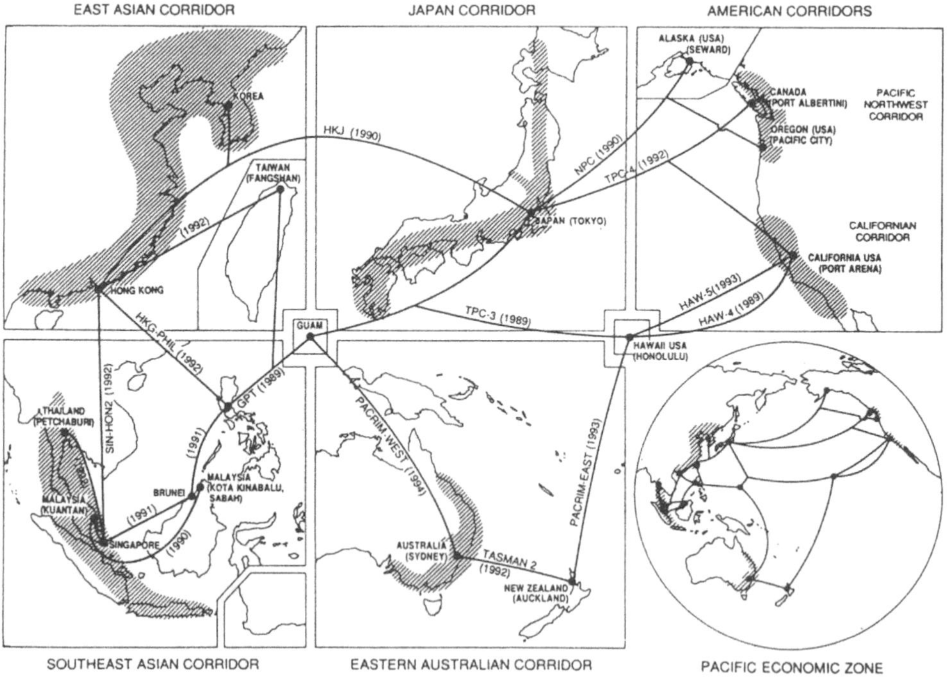

Figure 6. Network of fiber-optic cables in the Pacific Rim. Source: Rimmer 1991.

Dynamic coordination of logistical networks in the Pacific Rim should be sustained by the creation of well-coordinated, multi-layered, multi-modal transport and communications infrastructure. The network hubs are benefiting from being national and international focal points. They are trying to capture potential synergies and collective knowledge inherent in the dominant communications and transportation networks. The driving forces having been far beyond one nation's controllability are working to integrate nation's knowledge networks to global ones. The implication for national planners is that unpredictability and uncontrollability must be regarded as the essential factors for planning inquiries; that the international public goods, like international universities, international airports and the like are the major devices to accelerate and accommodate the global integration processes of knowledge networks. The past preoccupation with short-term futures based on classical principles of planning disciplines has to be abandoned in favor of long-term strategic planning which focuses on the qualitative features on the nonlinear dynamics of the integration processes.

TOWARDS A NEW PLANNING PARADIGM

In the tradition of neoclassical economics, technologies are supposed to be chosen by the cost minimizing (profit maximizing) behavior of individuals, firms and organization. The rationality underlying the human behavior is the very significant factor to understand the reality of technological change. By applying this analytical strand, we can predict, to some degree, the future states of technologies in our society. On the other hand, we can say that the rationality can play no decisive roles in our trial to predict the future states of our knowledge. We lack of the way to predict the future states of knowledge, simply because there is no way to know now what we can only know tomorrow. Thus, in time of innovation, in time of global change, in time of creativity, one must expect that the unexpected can occur; one must recognize that the emergence of the unexpected can ensure new opportunities and creativity.

The planning inquiries in the new arena should be based on a schema that can be modified by expansion as the planners and analysts learn more about the reality, and especially about the reality of the development dynamics of our society. The schema tries to depict societal systems in terms of decision makers who have a capability of changing certain things in our world. However, different from the planning paradigm in the industrial arena, the ends of new planning inquiries are to be identified no longer in terms of the desirable states to which decision makers want to alter their society, but rather in terms of the statement of his understanding about the reality of the development of social systems. Thus, the important feature of planning is that it is committed to ascertaining not simply whether the decision maker's choices lead to his desired ends, but whether they leads to ends that are ethically desirable for the present as well as the future and past generations.

Nowadays, it is extremely important to understand the incomprehensiveness in our trials to be comprehensive in the planning. We lack of the comprehensive knowledge about our society and human affairs. The planning paradigm, however, has not dropped its basic idea of to be "scientific", because the idea of comprehensiveness of planning inquiry is in effect exactly the same in intent as the classic approach to science, in which it was expected that the scientific effort could tie together in a meaningful way the various facts of natures and society. Today, in planning its adherents see no reason why all matters of human concerns should not tie together in one grand framework of a decision maker. All our goals can be tied together in a meaningful schema, and we do not have to resort to mysticism or some deep inexplicable human essence in order to describe them adequately.

REFERENCES

Andersson, Å.E., 1986, Presidential Address: The Fourth Logistical Revolution, *Papers of Regional Science Association,* Vol. 59, pp. 1-12.

Batten, D. and B. Johansson, 1987, Dynamics of product substitution, *Infrastructure and Building Sector Studies:* 11, CERUM, University of Umeå, Sweden.

Batten, D.F., Kobayashi, K., and Andersson, Å.E., 1988, Knowledge, Nodes, and Networks, - An Analytical Perspective-, In Å.E. Andersson, D.F. Batten, and C. Karlsson, eds. *Knowledge and Industrial Organization,* Springer Verlag, Berlin, Heidelberg.

Churchman, C. W., 1979, *The Systems Approach and Its Enemies,*

Basic Books, Inc., Publishers, New York.
National Land Agency, 1987, *The Fourth Comprehensive National Development Plan*, Government of Japan.
Rimmer, P.J., 1991, The emerging infrastructural arena in pacific Asia since the early 1970s, *Proc. of JSCE*, Vol. 431, pp. 1-17.
Yoshikawa, K., 1991, Future transportation and communication – A Japanese perspective, Paper presented at the 150 Anniversary Jubilee Conference of Swedish Road Administration, Borlänge.

THE FUTURE OF EUROPEAN TRANSPORT

Kenneth Button
Director of the Applied Microeconomics Research Group
Longhborough University
United Kingdom

INTRODUCTION

Making predictions is always dangerous and with respect to transport it can be especially so. The demand for transport services is derived from demands for goods and services and from the activity patterns of individuals. These are very difficult to forecast in their own right and, in addition, their interaction with transport is far from stable over time. Additionally, transport markets are seldom allowed to operate freely but are subjected to considerable political intervention. To attempt predictions concerning the future of European transport, therefore, requires making assumptions about wider economic, industrial and political trends. This means that forecasting future developments must inevitably take into account political and social attitudes as well as more technical considerations. In many ways, however, the early 1990s are already emerging as being watersheds in so many relevant areas that this has become a particularly difficult task.

In the mid-1980s I, together with Dr. David Gillingwater offered some general thoughts on the way transport policy might develop as we move towards the end of the century (Button and Gillingwater, 1986) but at the time of writing, which was only six or seven years ago, the idea of a single European market was only just being conceived, most Eastern European countries were still controlled by centralised Communist regimes and it was uncertain just how much further the Western, democratic economies would go down the road they had already embarked upon involving increased privatisation and extensive withdrawals of the state from economic intervention. Equally, while the natural environment was not ignored in the book – the concern with global and transboundary pollution as well as local environmental degradation was then apparent – the nature and extent to which environmental policy would be integrated into the long term formulation of political attitudes was still far from clear.

We now find, however, that economic, and to a somewhat lesser extent, political integration within Western Europe is both accelerating within the European Communities (EC) and is likely to spread geographically as additional states seek membership over the next few years. The unification of Germany is a further dimension is this process. Indeed, from another perspective, German unification may be seen as the extreme example of the on-going events in Eastern Europe where not only is possible membership or, more probable in the short term, association of post-Communist states likely to impact on the development of the EC, but internal economic and political changes will of themselves have major transport implications (Button, 1991a). We also now find that the natural environment is an ever present factor in the new political agenda and issues of excessive traffic congestion and global atmospheric pollution in particular are increasingly attracting the attention of transport policy makers.

All these factors make predicting the future of European transport difficult even in the context of the medium terms of ten to fifteen years over which we can make the reasonable assumption that no new technology will significantly impact on the situation. This does not mean, of course, that existing technology will not be developed nor that its use, especially in the informatics and telematics fields, will not be considerably extended.

Further, the time frame chosen means that several other parameters which have a major bearing on the demands for transport services can be taken as constant. In particular land use patterns and the scale and location of the main urban centres tend to change only slowing and it is these factors which have a considerable bearing not only on trip generation and attraction rates but also the distribution of much traffic. There will be movements of economic activities in Europe, both as macro-economic forces and political reforms cause shifts in population and as, at the micro level, the on-going trend towards greater urbanisation seems certain to continue while agricultural policies in both East and West Europe are reformed. Given the overall geographical size of the area, however, such changes are marginal to the overall picture.

We also have the problem of separating out what is likely to happen from what we would wish to happen. Transport is, for reasons perhaps best explained by psychologists, a topic which stimulates a considerable emotional response from many people. Indeed, it is almost inevitable that I have my own prejudices and biases with regard to how I would like to see European transport develop. Hopefully, one can separate out these normative views and offer an objective assessment of what is more likely to materialise.

The emotive factor does, though, introduce a second problem. If one looks at transport one finds that peoples' behaviour regarding both its supply and use does not conform to many of the normal assumptions of social scientists. Since deregulation in the USA. for example, we have witnessed individuals invest in airlines which earn and, must be expected on any reasonable criteria, to continue to earn an extremely low rate of return (Kahn, 1988). On the user side, trips are often made along anything but the minimum cost paths. There are, in other words, costs an benefits in transport which are not usually included in the conventional calculus of economics. This means that events and policies on a larger scale are often governed by factors well outside of the normal realms of predictive models, with their implicit assumptions of "rational economic man", employed by economists.

With these problems at the forefront my mind, in this paper, I attempt, in as neutral a way as possible, to highlight current on-going developments in European transport policy together with those which, while not yet implemented, are already in train and are likely to have longer term effects on transport in Europe. Linked to this will be a review of changes in the broader political and economic environment in which transport is supplied and demanded and the implications for change on transport. I restrict my assessment to an horizon of a decade or so to keep it within a range which is manageable and minimises the amount of speculation. The high costs, together with the substantial lead times, involved in introducing new technologies on a substantive scale mean that in taking such an horizon I avoid making what would be speculative assumptions about major technological changes.

BACKGROUND: THE CURRENT TRENDS

The future is never independent of the present and, in particular, in a dynamic sense changes which are taking place now inevitably have carry through effects. Furthermore, momentum is often inherent in the ways policies develop and future initiatives frequently follow on from policies now being implemented. Consequently, while historically one can point to exceptions, transport policy is, in this sense, generally more evolutionary than revolutionary. Realism, therefore, means that in looking at the future of European transport we should not be engaging in an exercise of comparative statics, whereby we start from an equilibrium position and seek to define a new equilibrium a decade or so. Rather, the situation is dynamic and fluid. In the context of European transport a number of on-going events will continue to influence trends in transport at least into the immediate future if not always to the end of the century. Some of these are institutional inspired developments while others are driven rather more by market forces.

There is already an on-going trend towards liberalisation and greater freedom of international transport within Europe. At one level, the *Cockfield Report* (1985) setting out the requirements for an internal EC market, highlighted the need to remove a number of specific transport related barriers to trade between the twelve members of the EC. Significant progress has been made in this regard although some of the new measures are still being phased in. The move to the Single European Market within the EC, for instances, will mean that by 1992 the documentation required to carry goods acress national bounaries will be simpler, the number of licences for international road haulage will be expanded so that is no longer constitutes a quantitative constraint on internal traffics, quantity and fare controls for international aviation within Europe will be relaxed and standardised technical entry requirements for many transport markets will have removed many qualitative constraints on operations and market entry (Button, 1991b). These changes hardly herald a completely free market but the direction of change is transparent and there is demonstrable momentum behind it.

Other initiatives at institutional integration have been developed outside of the strict Single European Market initiative, although some still comes within the broader scope of EC transport policy. Many of these on-going efforts extend well beyond the boundaries of the twelve. Several have involved efforts at coordinating investment in transport infrastructure of European-wide significance (Button, 1990). The EC devised a medium-term infrastructure investment programme i the mid-1980s (Commission of the European Communities, 1986a) and, although subsequently considerably watered down, has resulted in a programme of work of European significance to be carried through into the late 1990s.

Similarly, the EC in conjunction with other bodies, such as the European Civil Aviation Conference, has brought forward a package of measures to coordinate investment in air traffic control and navigation equipment on a European wide basis and have also attempted, less successfully, to stimulate the creation of a European high-speed rail network (Commission of the European Communities, 1986b). Again the planned time span of implementation extends into the late 1990s and beyond.

At the private sector level, significant changes in the way transport is viewed within the overall production process has meant that international freight transport has been going through a considerable period of change. In particular, integrated transport has grown in importance as companies have begun to seek the full efficiency gains associated with multimodal operations. Equally, reaping maximum economis of scale, density and scope are now seen as important ways for minimising costs of transport supply and the international transport supply industry is increasingly becoming globalised with megacarriers emerging in many sectors of both freight and passenger transport. It seems unlikely that either the full range of cost saving gained from multimodal activities or from multinational operations have yet been reached and the process must, therefore, be seen as an on-going one.

Coupled with significant changes in international transport are on-going developments at the national level. Over the past decade there has been a marked move towards privatisation of supply, liberalisation of markets and decentralisation of decision-making of transport policy in many European countries (van Gent and Nijkamp, 1991). The UK is perhaps the extreme example of this with deregulation of many transport sectors and privatisation of such things as bus services (outside of London) and airlines and quite substantial amounts of transport infrastructure (e.g. airports, sea ports and river crossings). Other countries have, to a somewhat lesser degree, also followed similar patterns (Organisation for Economic Cooperation and Development, 1990). By way of illustration, Sweden has began experimenting with separate rail track administration from rail operations and has been deregulating road haulage since the 1960s, Portugal has deregulated local bus services, France has ended its obligatory rate structure for road haulage and Ireland has deregulated aviation. France is also intending to withdraw subsidies form its trunk rail services.

Liberalisation of national transport is also gradually emerging in Eastern European countries as privatisation of sectors such as road haulage takes place and foreign investment is drawn into areas of infrastructure provision and the production of motor vehicles, aircraft and other mobile transport capital (Button, 1991a).

The overall underlying philosophy of national transport policies (and to some extent international policy) has also began to shift. Where as much of the concern of the policy-makers of the 1940s through to the early 1980s was on providing access and mobility (generally through ever increasing levels of infrastructure provision, protection of preferred operators, and the granting of operational subsidies), the appreciation of wider costs of this type of strategy have gradually produced something of a reappraisal.

On the economic side, questions of simple efficiency have been raised and this, in part, has led to some of the privatisation and reductions in subsidies cites above. In addition to this the environmental costs of transport have brought forth a variety of measures to contain its natural detrimental impacts and also to reverse the adverse consequences of some long established government policies (Button, 1991c). The former embrace such things as differential taxation on leaded and unleaded fuel, the fitting of catalytic converters, traffic calming measures in urban areas, noise limits on aircraft, regulations on maritime discharges, changes in tax regimes, etc. Some of these measures represent EC or Organisation for Economic Cooperation and Development (OECD) wide initiatives while others are national or regional in their origin but the processes is an on-going one with many regulations still being phased in and others about to be introduced.

THE FORCES FOR CHANGE

There are a diversity of influences operating on the demand and supply of transport and which ultimately determine its form and use. In the context of the European situation one can isolate a number of key factors which will affect events well into the next century. Some of these represent new factors in the equation and were of limited, if any, significance a decade ago while others are on-going forces whose presence will persist and continue to shape European transport. The more important of these forces are summarised in Figure 1 together with some brief headings highlighting the broad areas in which we can anticipate developments in transport stemming from the impacts of these forces.

It should be noted that while many of the forces for change may be seen as "sticks" reflecting, for example, the inadequacies of transport provision in Europe, and hence represent negative pressures for change, others take the form of "carrots", such as the changes taking place in East Europe, and should be viewed as offering positive opportunities for innovation and adaptation. Equally, some of the forces are institutional and have their origins well outside of the narrow confines of transport – developments in the EC obviously fall into this category. Others, however, are of a more narrow, purely transport orientation and involve such matters as changes in the technology used and the increased difficulties being posed by serious capacity constraints and bottlenecks in the European transport networks. Equally, the likely responses, and thus the future characteristics of European transport, will in part come from within the transport sector itself as transport supply industries and users react to new stimuli but, also in part, it will stem from official policies which treat transport in a rather broader context.

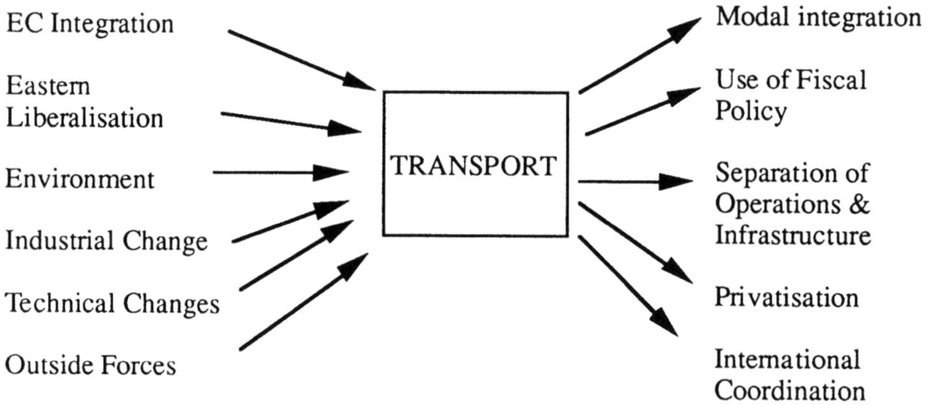

Figure 1. *The influences on transport developments in Europe in the 1990s.*

Finally, while for the sake of discussion it is convenient to categories, many of the forces for change and their ultimate impacts are, in fact, interactive and not strictly independent of one another. Indeed, given the political changes which are taking place both within Europe, and in the wider international economy, coupled with new approaches to industrial production these inter-relationships are, if anything, growing with time.

We now look at each of the main forces outlined in Figure 1; examining their current impact and possible future implications in turn. Following this we attempt to offer some comment on their implications for the future of European transport.

The European Communities Dimension

The probable impacts of the creation of the Single European Market after 1992 have attracted a lot of attention and certainly the new situation has impliations for transport policy both within the Twelve and with regard to a number of outside countries, such as Austria, Yugoslavia and Switzerland, which offer transit routes for Community traffic. The integration of European aviation within wider geographical markets also means that liberalisation of the EC policy has produced important knock-on effects across the entire European air network (Button, 1991d).

How many of the changes taking place strictly have to do with the Single market *per se* and how many with other, on-going pressures to develop a true EC Common Transport Policy as explicitly set down in the Treaty of Rome, is difficult to decide[1]. In terms of general transport policy, movement forward was forced on the EC Commission and Council by a ruling of the European Court of Justice in 1985 stating that the progress made in developing a Common Transport Policy had, to that time, been inadequate. Further, important legal rulings on individual cases, such as the *Nouvelles Frontieres* and the *Ahmeed Saeed cases,* have been instrumental in shaping certain, specific elements of policy. Whatever the immediate reason, in material terms long standing thorny problems surrounding cabotage in road haulage are finally being overcome, quantity controls over international haulage and aviation are being removed, issues of consistency in

---

1 It is often not appreciated that the Single Market concept only covers some aspects of the Common Transport Policy which also embraces elements of state aid policy, improvement of railway financing, harmonisation in the road sector, infrastructure planning and investment.

taxation are being addressed, etc.

In many ways, however, these measures are unlikely to offer any real solution to the underlying long term problems of EC transport policy. The basic problem is that efficient transport, as well understood by the founding fathers of the ECSC and consequently EC, is a central ingredient of a dynamic and prosperous European economy[2]. The measures to date have removed many of the former institutional constraints which limited the efficient supply of transport services but have to date done little to ensure that the use and provison of transport infrastructure approaches optimality.

In particular, while changes are taking place, or are in the process of being initiated, which are likely to produce some increase in the overall efficiency of the use made of transport infrastructure, and further EC supported investments will provide some additional capacity, they will not be sufficient to meet the forecast growths in traffic. There are now, for instance, over 140 million cars in Western Europe. The increase in road traffic has been particularly substantial in recent years with, for example, car traffic growing (in terms of vehicle kilometres) by 85 percent in Italy, 82 percent in the UK and 79 percent in Italy between 1970 and 1987. Over the same period, goods movements by road have risen (in terms of vehicle kilometres) by 93 percent in France, 93 percent in Italy and 45 percent in the UK.

More importantly for the future, the forecasts are that, with the continuation of existing policies, road traffic growth will continue and in the case of, for example, Great Britain may rise by up to 142 percent between 1988 and 2025 (Department of Transport, 1989). Predictions made on a similar basis for the Netherlands suggest a 72 percent growth between 1986 and 2010. More aggregate studies of the EC have forecast significant growth in road haulage, expecially international traffic, as manufactured goods take an ever increasing share of the overall amount of goods transported (Gwilliam and Allport, 1982). Equally, if one looks at European aviation a similar picture emerges with traffic growing from 321.3 thousand million passengerkilometres in 1978 to 507.9 thousand million by 1988 and forecast, by the International Civil Aviation Organisation, to rise to some 850.0 thousand million by the end of the century. We also already find that of the 46 largest airports in Western Europe, 12 are currently operating at or around their physical

---

2 Article 3 of the Treaty of Rome states explicitly, "...the activities of the Community shall include. .... the adoption of a common policy in the sphere of transport".

capacity and a further 11 will, it is predicted, reach capacity by 1995 (see, Figure 2). Further, the EC and others have identified a serious of major bottle-necks in the European surface transport networks and if traffic grows as forecast this number will expand and the problems intensify (see, Figure 3).

The underlying problem is that under current policies the efficiency of the transport system will inevitably deteriorate in Europe as congestion develops. The implications of this are summarised in a recent report submitted to the EC Commission (Group Transport 2000 Plus, 1991), *viz*: "....a general deterioration in transport conditions due to inefficient use of the networks and the saturation of certain infrastructures (especially road and air). Also – albeit not so immediately noticeable – there is an on-going increase in the nuisance caused by transport. The culprit here is not so much network saturation as the actual increase in traffic".

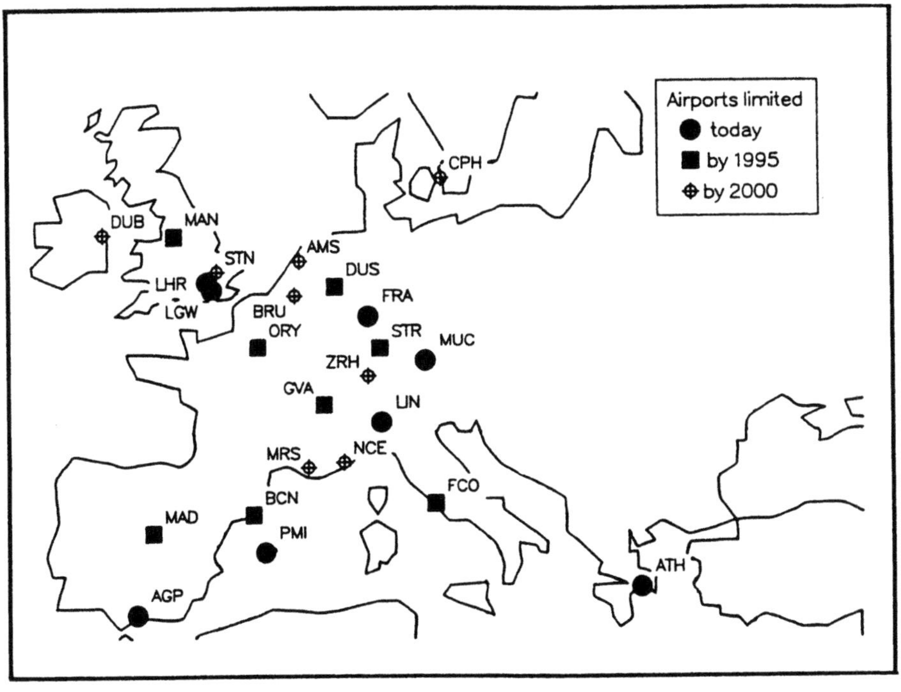

Figure 2. *Congested Airports in Europe. Source: Association of European Airlines (1987).*

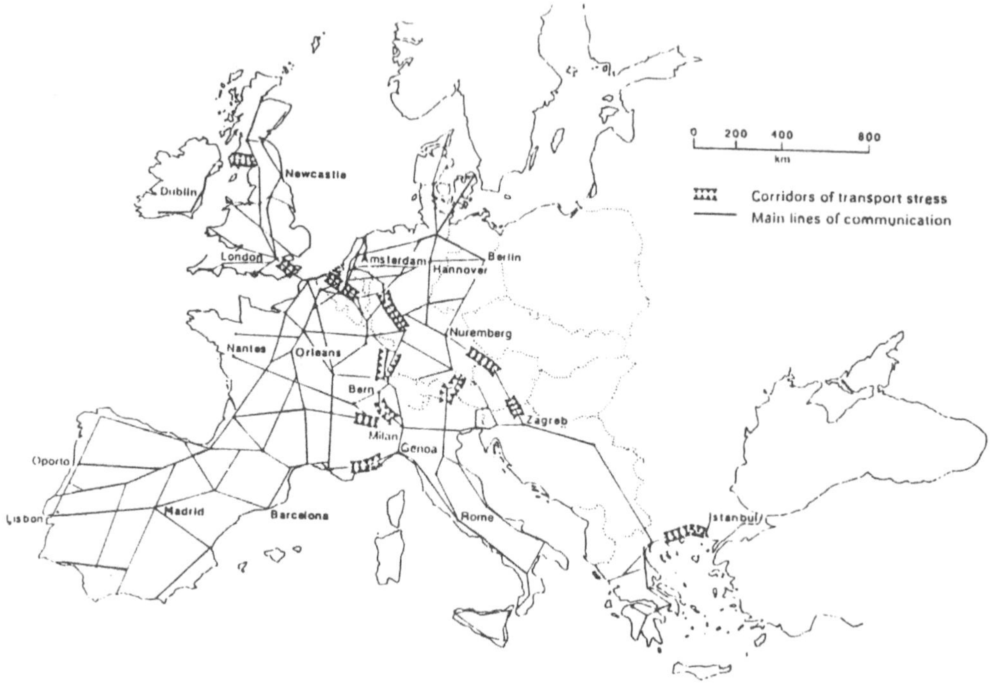

Figure 3. *Links in the European transport network experiencing stress. Source: Whitelegg (1988).*

Linked to this problem is the realisation that while there may be difficulties in terms of congestion of existing facilities, and as we see below massive expansions in capacity are unlikely to be the answer, there is a need to develop entirely new transport networks within in Europe (Roundtable of European Industrialists, 1990). This stems from both the changes taking place within the EC, its probable expansion of membership and the almost inevitable links which will be gradually created with the post-Communist states. There are forces, therefore, at work within the EC which indicate the concurrent need both for restraint measures to make better use of the transport infrastructure available in the face of rising demands and for the creation of new transport networks.

Eastern Europe

The liberalisation of Eastern Europe represents both problems and opportunities. It means in particular that the overall "transport market" in Europe will expand considerably in line with major new urban and industrial centres being brought within the market system. It means that many more major transport links must now be considered as part of Europe's transport future. In many ways this may prove advantageous in terms of the long term development of European transport since it creates something more akin to a natural market for transport services than currently exists.

Sort term problems, however, arise mainly because of the different attitudes regarding transport which has grown up in Eastern and Central Europe over the past forty years and the impact this has had on the physical transport infrastructure now in place. In particular, the transport systems of Eastern Europe are dominated by rail (which itself suffers from low productivity and over-manning), tend to be of poorer quality than in Western Europe and have been developed since the late 1940s to meet the trading patterns of the members of the Council for Mutual Economic Assistance (CMEA) (Button, 1991a).

Regarding trade patterns, the bulk of foreign trade has been within the confines of the CMEA with much less taking place with the rest of the world. In 1985, for example, Bulgaria exported (in $US terms) about five times as much to other CMEA countries as to the rest of the world combined and imported four times as much. In the case of Czechoslovakia the ratio was about 3.5:1 for both exports and imports and about 1.5:1 for the USSR and 2:1 for the GDR on a similar basis. These trade patterns inevitably led to infrastructure being designed to cope with the particular form of internal CMEA trade. Forecasts suggest that on the basis of even a relatively conservative scenario, by the early part of the next century this pattern of trading activity will be transformed and with this will come the need for a different form av transport provision (Seidenfus, 1991).

Additionally, the trade which does exist between the former CMEA countries and the EC, in particular, while relatively balanced in financial terms is heavily bias in favour of bulk movements from Eastern Europe in physical terms. In the context of aggregate tonnage carried by rail, while 30 million tonnes of freight moved from East to West only 15 million tonnes moved in the reverse direction in 1987. The comparable figures for road transport are 9.8 million tonnes and 3 million tonnes respectively. Within the former CMEA imbalances in the nature of trade also exist and, in particular, the USSR fostered the establishment of a set of freight rates which effectively subsidised the exports of its raw materials to other Eastern European states. Other countries benefited in that their import costs for raw materials were thus kept down but they also had to introduce similar pricing structures for their own transport facilities which meant that for through traffic the transited country often suffered heavy losses.

Problems of this kind also extend to the cost side with differences, in this case, in rail gauge between Poland and the USSR resulting in vast amounts of resources being expended on building axle-changing and reloading facilities at border points (Alvstam et al., 1985). These types of trade imbalances and technical incompatibilities, brought about primarily through lack of appropriate market incentives, resulted in considerable amounts of empty running coupled with investments in facilities ill-suited to the carriage of manufacturer products which accompanies economic development.

The Eastern European states are, especially in comparison with EC members, also heavily reliant upon rail for freight transport (Table 1) but, in addition, their systems tend to be old and the efficiency of operations is highly questionable. If one considers Table 2, for instance, it is clear that the Eastern European economies have extensive rail networks however, when compared to EC states, a relatively small percentage of them are doubled tracked. It is also clear that there is a need for more electrification. For example, of 27,284 kilometres of line, 11,669 are currently electrified in the states of pre-1990 FRG but out of 14,024 kilometres in the old GDR only 3,475 kilometres were electrified and in Poland, of 26,545 kilometres only 6,296 are electrified. The rolling stock, especially that required for passenger services, of the Eastern European railways tends to be older than that in the EC and also much more uniform which makes it difficult to respond to changing economic conditions (Timar, 1991).

In terms of productive inputs, the transport systems of most Eastern European countries are much more labour intensive than those found in the West. For example, while the pre-unification FRG's rail system employed 240,806 people in 1987, the GDR's employed 249,964 and Poland's 360,015 but in terms of performance, the FRG system did over 40 percent more tonne kilometres than the GDR's system and only 20 percent less than Poland's. In the USSR, which has the world's busiest rail system, nearly four million people are employed on the network, although, for comparative purposes, it is interesting to note that the US system, while admittedly producing only half the tonnage and a quarter of the tonne kilometres, operates with only about 150,000 employees.

Table 1. Mode split for freight carriage (thousand tonnes/kilometres in 1988)

Country	Road	Rail	Inland Water-ways
European Community			
Belgium	26.00	7.00	6.35
Denmark	9.10	1.66	-
France	111.80	52.30	7.33
Germany	151.40	58.50	52.80
Greece	14.00	0.60	-
Ireland	5.00	0.55	-
Italy	169.00	19.60	0.14
Luxemburg	0.40	0.64	0.36
Netherlands	22.10	3.20	33.85
Portugal	9.00	1.56	-
Spain	133.00	12.00	-
United Kingdom	124.80	18.20	2.30
Eastern Europe			
Bulgaria	14.70	17.60	-
Czechoslovakia	13.10	69.40	5.20
GDR	16.40	59.40	2.50
Hungary	13.20	20.20	2.00
Poland	38.80	120.70	1.40
Yugoslavia	20.90*	25.40	4.60

* Public transport only

Source: Taken from United Nations, European Conference of Ministers of Transport, International Road Federation and International Union of Railways data sets.

Part of the problem must be attributed to the lack of power management has enjoyed over recruitment and sacking but equally management has operated under a confused objective/reward structure which has done little to stimulate its efficiency or to motivate innovation (Ambler et al., 1985).

The road systems in Eastern Europe are not much better. The data in Table 2 provides some quantitative indicators of the kilometres of motorway in CMEA and EC states, but quality is also

Table 2. *Basic transport infrastructure of selected European countries (1988)*

Country	Motorways (kms)	Rail Track (kms) (double tracked)		Car per Capita
European Community				
Belgium	1570	3554	(74%)	0.364
Denmark	835	2476	(33%)	0.320
France	5300	34365	(45%)	0.400
Germany	8715	27278	(50%)	0.472
Greece	90	1565	(16%)	0.150
Ireland	8	1944	(27%)	0.208
Italy	6091	16015	(35%)	0.430
Luxemburg	75	272	(60%)	0.460
Netherlands	2060	2828	(63%)	0.356
Portugal	211	2850	(15%)	0.210
Spain	2344	12531	(21%)	0.276
United Kingdom	3100	16599	(69%)	0.349
Eastern Europe				
Bulgaria	258	4055	(23%)	0.132
Czechoslovakia	518	12856	(23%)	-
GDR	1855	14024	(30%)	0.225
Hungary	324	7403	(15%)	0.169
Poland	220	26545	(34%)	0.119
Yugoslavia	828	9349	(12%)	0.131

Source: Taken from United Nations, European Conference of Ministers of Transport, International Road Federation and International Union of Railways data sets.

important. In this respect the United Nations Commission for Europe feels that, "The present road infrastructure in Central/ Eastern European countries has deteriorated such that it is unlikely to be able to cope with increased road traffic volumes" (Lehmacher, 1990).

There is already a momentum at work to increase investment in improved infrastructure in Eastern Europe. Germany is putting resources into the up-grading of the former GDR rail network, although the costs appear to be massive. One estimate, expressed in May 1990 by Dr. Reiner Gohlke, president of the Bundesbahn at a conference of the Community of European Railways, is that some DM100 billion is needed simply to get the old GDR rail system into a comparable condition to the network in the West. The European Bank for Reconstruction and Development (EBRD) sees the improvement of the transport infrastructure of the post-Communist states as one of its main functions and indications are that it is going to commit a substantial part of its own resources to this end.

The Environmental Issue

Public concern over the state of the environment has unquestionably grown over the past decade and all the signs point to this concern continuing to influence policy making into the foreseeable future. This is not surprising, transport is a major contributor to local environmental degradation. From the traditional environmental economics side, for example, Baumol and Oates (1979) in their seminal work state that, ".... some of the most pervasive externalities are generated not by industrial operations, but by individual activities. The automobile is a notorious producer of detrimental spillovers. Automobile exhausts increase laundry expenses, make it more difficult to breathe, and even shorten lives. The heavy cloud of pollutants that hangs over crowded roadways, the widespread traffic delays, and the heavy accident rates in such areas again suggest that externalities created by the activities of consumers are not the exceptional phenomena."

Further, it is now widely appreciated that transport impinges on the environment in several important ways not considered a decade or so ago. In particular, it contributes, not only to local environmental problems of noise, atmospheric lead pollution, low level ozone concentrations, vibration, visual intrusion, community severance and particulate emissions but also significantly adds to transboundary pollution through emissions of nitrogen oxide (NO_x) and to global environmental degradation through its contribution to carbon dioxide (CO_2) emissions (Linster, 1990 and Transnet, 1990). There are other external costs associated with accidents and safety, especially with regard to non-motorists, although the existence of insurance markets often makes it difficult to disentangle the strict externality element. All modes are environmentally intrusive but the nature and scale of their impacts differ (see Table 3 for some comparisons).

Table 3. Environmental impacts of various transport modes

Mode	Air	Water Resources	Land Resources	Solid Waste	Noise	Accident Risk	Other Impacts
Rail			Land taken for rights of way & terminals; derelection of obsolete facilities	Abandon-ed lines, equipment & rolling stock	Noise & vibration around terminals & along lines	Derailment or collision of freight carrying hazardous substances	Partition or destruction of neigh-bourhoods, farmland & wild life habitats
Road	Local (CO, HC, NOx, fuel addi-tives such as lead & particula-tes) Global (CO_2, CFC)	Pollution of surface water & ground-water by surface run-off; modifi-cation of water systems by road building	Land taken for infra-structure; extraction of road building materials	Abandon-ed spoil tips & rubble from road works; road vehicles withdrawn from service; waste oil	Noise & vibration from cars, motor-cycles & lorries in cities & along main roads	Deaths, injuries & property damage from accidents; risk from transport of hazardous substances; risk of structural failure in old or worn road facilities	Partition or destruction of neigh-bourhoods, farmland & wild life habitats; congestion
Air	Air Pollution	Modifica-tion of water tables, river courses, & field drain-age in air-port const-ruction	Land taken for infra-structures; dereliction of obsolete sites	Scrapped aircraft	Noise around airports		Congestion on accesss routes to airports
Marine & Inland Water		Modifica-tion of wa-ter system during port construc-tion & ca-nal cutting & dredging	Land taken for infras-tructure; dereliction of obsolete port faciliti-es & canals	Vessels & craft withdrawn from service		Bulk transport of fuels & hazardous substances	

Source: Linster (1990).

Table 4. *NO_x emissions from transport (10^3 tonnes per annum) in Eastern Europe*

Country	All Modes 1980	Road Transport 1980	Road Transport Forecast 2000
Bulgaria	89	88	185
Czechoslovakia	91	88	95
GDR	-	81	144
Hungary	100	75	142
Poland	190	185	201
Romania	-	85	105

Source: Laikin et al. (1987).

Transport induced pressures on the natural environment are almost inevitably going to become more intense over the next decade without significant policy changes for at least three interlinked reasons.

First, in physical terms the sheer scale of transport activities explains part of their impact on the environment and especially, in the context of global climate change, their level of CO_2 emissions. There are now over 125 million goods vehicles in the world and over 400 million cars and, as shown earlier, many of these are in Europe. The projections cited above indicate there will be a continued increase in traffic volumes in the industrialized world and previous experience suggests that, while individual vehicles can be made more environmentally friendly, the sheer scale of this growth will lead to considerably more environmental degradation (Barde and Button, 1990). In the UK, for instance, we find that while cars are now about 20 percent more fuel efficient than a decade ago, and hence individually less CO_2 is emitted, there are now over 35 percent more cars on the roads with car traffic up by over 40 percent (in terms of passenger kilometres) over the same period.

Growth in traffic is not only likely to occur under existing policies in Western Europe but will also be seen in the post-Communist states as their economies increase in efficiency and disposable income rises. One could argue that as the Trabants and Ladas are replaced by more modern, less environmentally damaging vehicles so there will be some containment of the problem. In fact, this is unlikely to have any significant impact

Table 5. *Forecasts of car ownership in low income countries (1986=100)*

Country	Annual Rate of Per Capita Income Growth	Forecast 1986	National 2000	Car Park 2025
Burkina Faso	1%	100	143	286
	4%	100	183	394
Rwanda	1%	100	177	499
	4%	100	224	951
Togo	1%	100	149	335
	4%	100	205	751
Haiti	1%	100	96	161
	4%	100	128	351
Pakistan	1%	100	148	340
	4%	100	196	739
Cameroon	1%	100	162	440
	4%	100	255	1309
Gabon	1%	100	215	632
	4%	100	355	1922
Algeria	1%	100	141	341
	4%	100	232	1000
Mauritius	1%	100	146	209
	4%	100	219	544
Malaysia	1%	100	149	284
	4%	100	216	628

Source: Button and Ngoe (1991).

when set against the release of current material aspirations of the populations of Eastern and Central European countries to enjoy the personal benefits of private transport. Forecasts are very crude as to exactly what the growth of car use will be in these countries and of the environmental implications are even less developed. We can seen from Table 4, however, that the tentative predictions which have been made of NO_x emissions suggest that unrestrained traffic growth will add, at the very least, to existing transboundary pollution problems.

The third element concerns events outside of Europe and, in particular, in low income nations. If incomes rise and economic activity expands in the way the World Bank and other international agencies anticipate, then it is forecast that car and ownership and traffic in less developed countries will rise even more dramatically during the next 35 years than is predicted for Europe. Getting reliable information on even the existing car parks of low income countries is difficult and any predictions about the future must be surrounded with caveats. However, deploying the types of forecasting models used in industrial countries predictions can be made (Button and Ngoe, 1991). As can be seen from a sample of results in Table 5, the indications are that vehicle ownership, and particularly at the higher end of the spectrum of low income countries, will increase substantially over the coming decades assuming existing policies remain in place. The relevance of this for Europe is that, on the assumption that there is no intention of forcibly restricting these countries in their desire to enjoy the material living standards now available in Europe, there will be a need for Europe to significantly reduce the global environmental impacts of its transport even if a steady state situation regarding CO_2 emissions is to be achieved.

Industrial Change

There have been significant changes over the past decade in both the make up of industry in Europe and in the way that industrial production (in an overall sense) is conducted.

In terms of composition, there has been a relative switch within Western Europe to the manufacture of final goods rather than the production of intermediate goods or the extraction of raw materials. But even within manufactures there has been a relative shift away from more traditional goods to high-technology products. Goods, therefore, tend to be less heavy and to be moved directly to warehouses or customers rather than from plant to plant. The implications, and these emerge from the basic data, has been a significant increase use in the use of road freight transport. Between 1970 and 1989, for instance, rail freight (in terms of tonne-kilometres) fell by over 4 percent in European Conference of Ministers of Transport states while road freight rose by over 110 percent. Less than 10 percent of international freight traffic within the EC is now moved by rail. While it is possible that there may be changes in the future affecting the structure of Western European industry which will favour the use of rail, the indications are that these are unlikely and industry will continue to seek more of the attricutes offered by road transport.

The composition of the European labour force has also changed quite radically over the past two decades with both significantly higher levels of female participation in most countries and more part time work being done. This, in turn, has affected the patterns of journeys of work undertaken in urban areas, leading in particular to more emphasis on car use, and has influenced the mix of work/leisure travel. While there are national exceptions, changes of this type have now largely run their course, and hence although travel behaviour will continue to be influenced by forces associated with changing labour participation rates, the major forces for change have now probably already been felt.

It is not just the nature of the goods produced which is impacting on European transport, it is also the changing role that transport is playing in the manufacturing process. The use of just-in-time (JIT) management approaches have expanded throughout the industrialised world (Schneider, 1985). This essentially involves producers minimising inventories of finished goods and inputs by operating transport as an integrated part of the production process[3]. To ensure both a steady inflow of components and of raw materials combined with rapid dispatch of final output to customers requires reliable and fast transport. Simple financial cost minimisation, therefore, gives way to trade-offs of cost against speed and reliability.

Exactly how widely JIT is employed by European industry is difficult to say. Certainly asking producers seems to shed little light on the question with many claiming to have fully integrated JIT systems but holding demonstrably high inventories while others, claiming little if any knowledge of the textbook JIT model, appear to have intuitively adopted it (Button, 1989). Scanning the literature suggests that in the UK, which is certainly not the most innovative European country in this area, about 10 to 15 percent of firms employ effective JIT systems but this compares with over 40 percent in the USA and 70 percent in Japan.

The point of these figures is that, despite their crudity, they do suggest that many Europe countries have some way to go to catch up with their international rivals in the use of JIT techniques. Catching up will inevitably have implications for transport supply,

---

[3] The adoption of JIT techniques does not just affect the way production is carried on but often requires new manufacturing hardware and this has long term implications for such things as industrial location patterns. As Drucker (1990) says, with JIT, ".. the plant no longer functions as a step-by-step process that begins at the receiving dock and ends when finished goods move into the shipping room. Instead, the plant must be redesigned from the end backwards as an integrated flow".

and especially mode-split, over the next decade. The attributes sought by companies practicing JIT techniques suggest that pressures will be towards supplying appropriate road and air services rather than rail or inland shipping which, while often cheaper in terms of pure transport costs tend, because of their lower speed and relative inflexibility, to tie up large quantities of inventories. The need for additional transhipment, with associated problems of potential losses and damage to goods, also counts against road and inland water transport.

Technical Developments

Forces for change are supply as well as demand driven and changes in the nature of available transport hardware and methods of operating influence peoples expectations about policies toward transport.

While it seems very improbable that there will be major technical breakthroughs in transport over the next decade, there are on-going technical developments which by virtue of their impact on costs seem likely to influence transport. In particular, there are significant advances in informatics and telematics, especially stimulated by various research initiatives within the EC (such as the DRIVE and PROMETHEUS programmes) which could well bring about changes in the efficiency of private transport. Systems providing better information on routing, traffic conditions and potential impedance (e.g. accidents, road repairs, etc) offer the scope for expanding the effective capacity of the overall European road network. Improved navigation and control systems have the same potential for aviation. Improved electronic data interchange (EDI) systems permit more efficient systems of JIT management of production together with facilitating greater efficiency in transport distribution.

There are also changes in telecommunications which have implications for the demand for travel more generally. Improved telecommunications, in particular, open up the prospect of teleshopping, teleconferencing, telecommuting, teleteaching, etc which may well have longer term implications for the patterns of personal travel. Some years ago there was a feeling that such new technologies would quite rapidly and dramatically reduce the demand for personal travel but subsequent research suggests the relationships between telecommunications and travel behaviour are more complex (Salomon, 1986). In some cases, for example, greater ease of arranging travel brought about by telecommunications actually seems to result in trip generation while, in others, the time freed by teleconferencing or teleshopping, for instance, is used for other forms of trip-making.

Perhaps a useful way to think of the impact of telecommunications on transport is to take a diagram developed by Nijkamp and Salomon (1989). Over time the demands for person interaction tend to grow. This seems to be partly a function of the growth in the service sector coupled with increased leisure opportunities for individuals. There, are however, constraints on the personal interaction. For one thing, people only have so much time available and, for another, constraints on available transport limit the number of trips which can be made. These facts act to place an upper limit (a saturation level) on personal interactions, albeit a flexible one in that technical change or increased transport supply offer scope for some upward movement. A sigmoid growth path of the general form depicted in Figure 4 seems to broadly map out the relevant path. What improved telecommunications potentially does in this context is (a) to push up the constraint in some instances where it facilitates more personal interactions for a given transport network and individual time budget and, more frequently, (b) to act as a substitute for travel and provide what is labelled the "telecommunications gap" enabling combined interactions, both personal and via electronic means, to follow the unconstrained path.

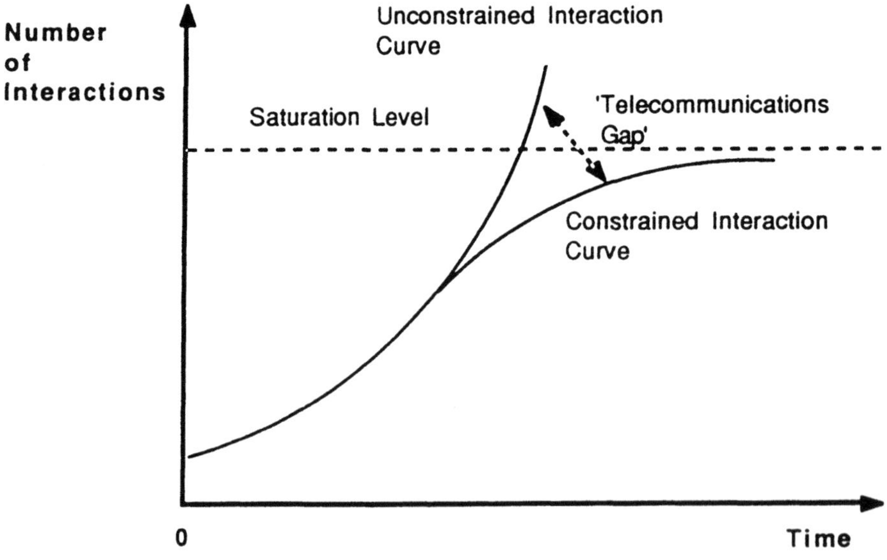

Figure 4. *The telecommunications gap.*

The pressures for change exerted by advances in telecommunications are likely to be felt as much in the communications sector as in transport. But from an overall policy perspective, and in terms of the impacts on travel behaviour, there are good reasons to believe that certain types of trip are more likely to be substituted by telecommunications alternatives than are others (Maggi, 1989). In particular, regular interactions involving quite simple interchanges of information are those most likely to be substituted especially if they are over long distances. Rising levels of traffic congestion in Europe and rapidly falling real costs of telecommunications hardware, coupled with a general increase in familiarity of electronic equipment, provide significant incentives for increased use of electronic media.

Outside Forces

Europe is not isolated from the rest of the world. It is heavily involved in international traffic to and from the Continent and in some sectors, therefore, its own policies cannot be insulated from these wider global markets. Equally, information on technical developments is soon disseminated and European transport systems have a tradition of adopting and, where necessary, adapting new technologies when they appear worthwhile. There is, however, another and equally important way in which transport in Europe is affected by what goes on elsewhere; this is on the demand side via demonstration effects regarding the impacts of transport policies and costs and benefits of various forms of travel and transport adopted elsewhere.

Looking first at the spread effects of changes elsewhere in international transport we already find that European aviation and shipping have been considerably affected by forces from outside. The "open-skies" policies adopted by the USA since the 1970s have lead to considerable liberalisations of the North Atlantic aviation market and this is gradually having knock-on effects on intra-European aviation. The UNCTAD attitude towards shipping conferences has also led to Europe accepted policies which implicitly limit the power of its liner shipping. These are forms of international pressures which, in various forms, are likely to continue to grow in importance as both international trade expands and the transport supply industries becomes more and more global in their orientation.

Regarding demonstration effects, these have proved important in the past not only in terms of stimulating technical changes and but also as catalysts for redirecting transport policy debates. On the technical side, the success of the Japanese "Bullet Trains", for example, laid the foundation for technical advances in high-speed railways in France. With respect to policy, the effects of liberalising US domestic aviation (Pickrell, 1990) has regularly been used as an argument in favour of freer transport markets in Europe. Equally, the events in the US since 1978 have provided demonstrations of where government interventions can most usefully be initiated when competitive (or contestable) forces fail to achieve their objectives. The ECAC and EC regulations regarding computer reservation systems (CRSs) systems and the EC Commission's attitude on mergers (Argyris, 1989) are illustrative of this. At a somewhat different level, much of the current debate in European countries over the adoption of road pricing to reduce urban traffic congestion stems form the success of the Singapore area licensing scheme and the experiment with electronic cordon pricing in Hong Kong.

Demonstration effects are important, therefore, both because they can set in train band-wagon effects and, in other instances, because they can provide support for one side of a policy debate already in progress. Their relevance for the future of European transport is, however, likely to be much less important than in the immediate past. This is quite simply because changes going-on elsewhere are relatively minor to many of those seen in the 1970s and 1980s. The implications of these earlier events, while still working their way through, are now largely discounted and, for example, do not of themselves provide further impetus for liberalisation.

THE EUROPEAN RESPONSE

The future of European transport in the medium term is likely to be shaped by the forces outlined above. What this implies is that changes are inevitably gradually going to be seen in both the way policy is formulated and the way that transport industries react to the new environment and challenges that will confront them. The outcome is almost certainly going to be an evolutionary change in the nature of European transport. We can think of the types of impacts which we observe under a number of separate headings. They are set out in no particular order of importance.

Modally Integrated Transport

Integrated transport systems of various kinds have been discussed for many years in Europe but policy-making has continued to be dominated by modal demarcations. This situation is, however, changing and is likely to continue to change in the future. The pressures of increasingly congested infrastructure and concern with environmental protection on the one side coupled with the needs of modern industry and the increased size of the potential European market on the other are leading to rather more careful thought as to where scarce marginal resources should be allocated. The implicit trade-offs involved mean that inevitably the issue of modal integration becomes important. The development of efficient EDI technology also means that one of the main problems with supplying high quality integrated transport services has been overcome.

In the past efforts at encouraging the use of such integrated systems as "piggy-back" and "kangaroo" trains have tended to rely on subsidies and manipulation of licensing arrangements[4]. Integration has also tended to be thought of in terms of technical possibilities and to be viewed as matters of national concern. The difficulties here are those of cost and of technical standardisation (Leijer and Tanja, 1989). In countries such as Switzerland geographical and environmental factors provide both commercial and social incentives to adopt combined transport which fosters the use of rail in particular. The large amounts of transit traffic add to the logic of this. In other cases, the nature of modern production and distribution suggests that other notions of integration are of more importance. These notions are being developed within the transport industry rather than policy-makers trying to impose them from outside.

The need for high quality and reliable service which, for example, is associated with JIT procedures, coupled with the demonstrable economies of scope and density which are associated with hub-and-spoke operations, is increasingly leading to more consolidation and transhipment. EDI systems now enable rapid discovery of problems in the system and facilitate detailed tracking of goods and vehicles. Modal integration is, and will increasingly, come via the market rather than through government inter-

---

[4] For example, within the EC countries such as Germany have traditionally given additional bilateral road haulage permits allowing other members' hauliers additional movements in the country if a designated number of trips were made on the combined transport facilities (Button, 1984).

ventions. We already see this happening with the very rapid growth of express parcel services in Europe (e.g. TNT, DHL, UPS, etc); a new sector relatively free of restrictions. These companies operate integrated long-distance road and air services and local distribution and collection via hub-and-spoke structure. On the passenger side, Lufthansa operate rail services to integrate with their air services. With regard freight, recent developments include shipping lines, such as Royal Nedlloyd Group, buying into port container facilities and road haulage companies.

While much of the current impetus for intermodalism is coming from new developments on the operations side, one can also point to important trends in infrastructure investment. The Channel Tunnel, while for rail use only, has necessitated the development of new lorry/rail loading techniques to cater for through road freight traffic. Airports are increasingly being treated as modal interchange points with the French TGV system, for instance, being linked to Paris (Charles de Gaulle) airport. The Swiss have mainline railway stations under their airports[5].

The position of the established intermodal carriers has been reactive to this situation. The Belgium Joint Declaration signed in 1990 by organisations (especially rail based ones) involved with international European intermodal transport, for example, has produced an agreement on a common strategy covering the design, operation and active marketing of intermodal transport. While such coordinated strategies have not produced much success in the past, the new market situation in Europe coupled with the emergence of new, mainly private sector players may well prove more stimulating for those involved in these initiatives in the future[6].

Modal integration is also likely to gain ground in a wider sense if one looks more broadly at transport and communications rather

---

5 Indeed, the aviation industry itself is increasingly, given airport congestion, coming to the view that many short distance legs of multi-leg trips are best served by rail air combinations. The Chairman of Guinness Peat Aviation, for instance, has stated, "High level political direction throughout Europe is needed to ensure that air and rail systems, currently being developed separately, are now developed as one" (Ryan, 1989).

6 The prospects for combined transport in Eastern Europe are somewhat less clear. At present little freight is even containerised — e.g. in the German Democratic Republic (GDR) prior to reunification only about 1.5 percent of traffic by volume went by container. There is, thus, scope for investment and expansion but at the same time it is not clear that the necessary framework for efficient decision-making exists and there may well be a tendency to put excessive resources into outmoded technology.

than focusing exclusively on the former. Telecommunications offer an important substitute for person trips in particular contexts. Also while the generalised costs of person movements can be expected to rise over time the telecommunications industry suggest that in some spheres, such as teleconferencing, the financial costs of interaction may fall by over 90 percent by the turn of the century. The adoption of common international standards for many forms of telecommunications equipment removes one of the major uncertainties for those contemplating purchasing hardware. There are also economies of scope to be enjoyed as the numbers integrated into telecommunications systems expand. The incentive in a society increasingly dominated by information requirements is, thus, likely to be greater integration of individual movements with telecommunications movements. Inertia, coupled with a certain fear felt by some people, is likely to restrain a rapid movement in this direction, but the pressures and possibilities are now such that a gradual integration of the way person movements and telecommunications are treated when decisions about interactions are taken is to be anticipated. The outcome will be not only that greater use of telecommunications will, of itself, act to constrain some of the otherwise forecast traffic growth, but also that it will act to buffer the potential transport users against the impacts of policies designed at traffic constraint.

Fiscal Reforms

One thing seems fairly clear, the overall costs borne by users of transport within Europe are going to increase in e short and medium term. This is quite simply because in the past transport has not covered its full costs and has enjoyed pricing structure which have not led to its efficient use. The circumstances allowing this are now changing.

Transport is an economic activity but has, in the past, tended to be treated rather differently to many other economic activities. This is one, many would say the main, reason why the transport networks of Western Europe are experiencing increased congestion and those of Eastern Europe are sub-standard. In the former case, prices paid for using transport infrastructure bear little resemblance to economic prices. Setting aside issues of environmental costs, the methods of charging for road, port and air transport infrastructure seldom bear any relation to the opportunity costs involved. In particular, users are simply not aware of the congestion that individual trips impose on others. The problem in Eastern and Central Europe has been that transport has not been treated as a "productive activity" and resources have been allocated so as to avoid duplication irrespective of potential demand. As such traffic and investment have been directed by administration rather than by market requirements.

As we have seen the pressures of increased transport are mounting in Western Europe and the old ideas of building more infrastructure have been seen as having limited scope for reducing the problem. Technical improvements, while helpful, are also not a long-term solution. More rational pricing of transport in Europe is increasingly been seen as the way forward. At one level there is obviously something of what, in the UK, is called the TINA ("There is no alternative") syndrome at work but additionally there is mounting recognition that appropriate economic pricing has the advantage not only of making more efficient use of existing infrastructure but also in giving clearer guidance as to where expansions in the system will generate the greatest benefits. It more closely links transport use to investment (i.e. "earmarking") priorities, and in helping to reduce the external, environmental costs associated with excessive congestion generates social benefits.

The containment of urban road traffic by means of "road pricing" is the most obvious example of where fiscal reforms could help to reduce congestion and excessive peaking of traffic. It is also an area where a number of European cities, most notably in the Netherlands, Sweden and the UK are actively reviewing change (Jones and Hervik, 1991 and Button, 1991d). While there is a very strong probability that reforms of this kind will gradual be introduced into many urban areas in Europe over the next decade, closer links between transport charges and infrastructure use are also likely to extend in other ways. The EC Commission, for instance, has in the past considered the possibility of economic pricing for airport use and, although rejected in the short term in favour of less political sensitive albeit less efficient alternatives, this has not been completely removed from the policy agenda.

The need to generate revenues for new infrastructure has also brought about innovative schemes which directly relate the costs of construction to those who will use the new facilities – toll rings in Bergen, Trondheim and Oslo to finance local road and other transport projects are examples. Plans for additional tolled, and sometimes private, motorways and trunk roads in the UK and several other European countries are additional illustrations

(Organisation for Economic Cooperation and Development, 1987).

One must add to these purely transport considerations the question of environmental costs. The mounting awareness of the environmental damage caused by transport, and in particular road transport, is creating political pressure for remedial actions. Some of these will inevitably come about through the introduction of regulations and standards, or the tightening and more stringent enforcement of existing laws. These actions, will not be costless in narrow economic terms, and despite their accompanying stimulus for greater efficiency, will lead to rate and fare and other price increases for transport use. Again the gap between payments and resource costs will be closer.

More transparently, there is mounting evidence that the exploitation of the physical controls available has very nearly reached its effective limit and that direct fiscal measures (e.g. pollution charges, tradeable emission permits, differential taxation, etc.) are now required. Many also argue they, in part because of their flexibility and revenue raising potential, are more effective tools anyway (Barde, 1991). The wider use of these policy instruments to restrict environmental degradation would directly bear on the perceived costs of transport.

Since neither congestion nor fiscal concerns on the part of macroeconomic policy makers are likely to diminish over the next decade, nor is the question of environmental quality going to disappear, one must anticipate increased emphasis on more soundly based policies regarding the funding and charging for the use of transport infrastructure. There is already evidence of considerble thought being given to the fiscal aspects of transport provision and operation in Europe, and this is being seen at all administrative levels, and the next decade will probably see rather more of this translated into actions than in the past.

Infrastructure Provision

There are at present major inconsistencies in the ways that the different modes of transport are treated within Europe. There is a clear separation of infrastructure ownership and control in aviation, shipping and with regard to road transport but, with limited exceptions, rail operations and infrastructure come under unitary authorities in all European countries. The logic of having unitary authorities for the railways stems from managerial notions of coordination coupled with technical arguments regarding the problems of separating operations from infrastructure. In particular, it is claimed that unlike air, road and sea transport there is only limited scope for overtaking on a railway which poses problems for timetabling and rescheduling in the event of late running or accidents if anything other than a fully integrated management system is adopted (Adamson et al., 1991).

Recently, however, concern over the efficiency of regulated, monopoly railway undertakings, especially regarding the ways subsidies are used, have raised questions regarding the desirability of separating responsibility for operating services from the responsibility for infrastructure (i.e. track, signalling and stations) management. This concern dovetails into problems of revenue generation for the maintenance and improvement of rail infrastructure.

While monopoly control of infrastructure, with associated problems of potential inefficiency, remain under this type of regime it has been argued that competition for use of this infrastructure by operating companies, both private and public, would help to enhance performance in at least this aspect of the overall rail system. Essentially a track authority would sell, possibly through a bidding process, "space" on its track, etc to operating companies. Further, if there was a European wide authority responsible for the main, internationally important infrastructure it would assist in coordinating investment programmes and help ensure greater compatibility of technical standards[7].

This divison of responsibilities is one option which has been adopted, albeit in a limited form, in Sweden and is under serious review in the UK. At the international level separation has also been suggested as a policy for improving international rail systems, i.e., "Any railway undertaking .. shall be granted access on equitable conditions to national railway infrastructure in the country of its establishment" (Commission of the European Communities, 1986b).

There are obvious difficulties with implementing a large scale divestiture of operations from the existing railway companies and with injecting competition into the use of rail infrastructure. The question is really one of degree but there seems to be increasingly strong arguments that some elements of the European rail system could be treated in this way (House of Lords Select Committee on the European Communities, 1990).

In terms of new transport infrastructure, the key question will

---

[7] There are major differences, for example, between the German and French high-speed rail networks which makes through running impossible.

be that of ensuring efficiency in the expenditure of resources. Technically, the faster one can provide a transport service the smaller the physical amount infrastructure which is required (de Waele, 1991). The difficulty with the vast majority of Europe's transport infrastructure is that it is multi-purpose and, therefore, tends to reduce overall traffic speeds with implications for the capacity required. There are exception, the TGV in France being the most obvious, but roads and railways in Europe carry both passenger and freight transport. They also often cater for local and long-distance traffics.

Specialisation of infrastructure therefore has considerable merit and there are also good reasons, in the case of railways, to focus investment on the major arteries[8]. Whether this will happen in the short term in Europe, however, is difficult to predict. Entrenched modal interests, coupled with ingrained notions that combined management of operation and control of infrastructure are essential in railways, tend to suggest that progess will be slow and partial. The pressures on the existing system, though, both in terms of simple capacity and the nature of the services users are seeking, are such that changes do appear almost inevitable.

Privatisation

Over the past decade there has been increased privatisation of transport in Europe (European Conference of Ministers of Transport, 1990 and Button, 1991d). This has involved some selling off, either in their entirety or in part by means of equity issues, of formerly nationally owned assets but more often it has involved the supplementation of nationally owned assets by additional private sector provision. Opening up formerly state monopoly markets for transport services to private sector suppliers is an example of the latter.

Much of the impetus for this has been generated by the need for more rapid expansions of infrastructure than national governments, for macroeconomic reasons, were prepared to finance from public funds. Additionally, some of it has come about because of concerns over the efficiency with which public enterprises are

---

[8] Indeed, one of the dangers in an area such as the former GDR is that resources will be put into up-grading too much of the secondary system. At the time of reunification there were some 14,000 kilometres of rail track but only 4,000 are required to link cities with populations of 100,000 or more.

operated[9]. Ideological considerations have also played their role.

From a predictive perspective it seems inevitable that private sector involvement in both the provision of transport infrastructure and in transport operations will increase. The reasons which have led recent trends are likely to strengthen rather than diminish in the future. Certainly it is difficult to see any reversal.

First, even if effective restraint measures stem the increase use of transport in Europe there are still bottle-necks which require investment and the maintenance of existing infrastructure is likely to pose increasing financial burdens. In particular, Eastern Europe requires its transport infrastructure to be modernised and its quality generally up-graded. It seems unlikely, to put it rather weakly, that this can all be funded from public monies. Indeed, the European Bank for Reconstruction and Development, which is to devote a considerable part of its investment funds to transport investments in Eastern Europe, still sees many of the necessary resources being drawn from the private sector.

Second, while there are national differences in attitude, the general trend in Europe is towards greater involvement of the private sector in supplying goods and services. The developments in the post-Communist countries means that this trend will, on an overall European basis, inevitably continue into the foreseeable future. Linked closely with this, the issue of efficiency, while not totally dominant, has become increasingly more important and there seems no reason why this will be reversed.

Third, as Europe becomes more economically (and possibly politically) integrated so national transport policies will have to be subsumed into wider international policies. National ownership and operation of transport has, whatever the initial intention, the inherent danger associated with it that national governments will attempt to manipulate it to its own economic ends. This has been one of the traditional stumbling blocks in the creation of an EC Common Transport Policy. Although not strictly related to privatisation, the efforts of the EC Commission to separate railway infrastructure from national ownership and also to put more resources into the Commission controlled element of the EC Transport Infrastructure Fund are early indications that this problem are being appreciated at the EC level. Taking this one step further, the removal of elements of transport supply from national or Commission ownership and giving it over to the private sector

---

9 The major problem being that of developing, in the absence of market criteria, an effective set of criteria for management to enable them to judge their performance against the general, policy objectives set them.

reduces political involvement further and could well prove more attractive to many Members.

International Coordination

The apparent consensus amongst forecasters is that, in relative terms, international transport within Western Europe will grow more rapidly than national transport over the next twenty years. This does, however, contain the implicit assumption that there will greater coordination of transport policies. If one simply looks back at the history of the Common Transport Policy of the EC this may seem a rather optimistic assumption. Agreeing on common objectives let alone policies for achieving them has proved extremely difficult and the attitude has continually been one of states both protecting their own transport supply industries and using transport policy as an instrument of national industrial advancement.

As indicated above, changes are now taking place as the EC becomes a Single Market and as Eastern and Central Europe begin to appreciate the need to integrate their markets with those in the West. To overcome some of the bottle-necks which already exist in the transport system, and others which are going to develop as patterns of trade change, European infrastructure investment and transport regulatory policy must inevitably become trans-national in its orientation. There, however, are inevitably going to be limits to the extent to which fully effective integration occurs.

It is now appreciated that there is a danger when attempting international coordination of transport to seek some "blue print" utopian plan of what the ultimate networks should look like. Experiences from the past, and especially regarding the early years of the EC Common Transport Policy, suggest that this approach to integration is both difficult to carry through (mainly because of the diverse views of the states involved as to the form these networks should take) and, in any case, given the lengthy construction periods involved, it is improbable that the outcome will in any way conform to final notions of the ideal system.

To be successful, integration must be strategic in its orientation and flexible in its implementation. What we are increasingly likely to witness in Europe in the foreseeable future is progress towards a development of such an approach rather than its full implementation. It is also, in a sense, likely to be reactive in its ethos. There is already agreement that key bottle-necks in the existing infrastructure should be removed, agreement on some aspects of controlling environmental degradation, agreements on some standards, formalised mechanisms for the interchange of information, etc. These are areas which are inevitably going to be built upon and developed and perhaps, given the complexity of the issues involved and the changes which are taking place, they are the best that can be hoped for.

CONCLUSIONS

Historians often talk of Industrial or Transport Revolutions but, with some notable exceptions, these are normally seen as spanning many decades and represent kinks in a development process rather than a completely new path being established. Certainly, over the time frame we have chosen there is unlikely to be even a minor kink let alone a major shift in the development path of European transport. What we can expect over the next ten to fifteen years, however, is evolution in a number of important areas brought on by the interplay of number of forces.

We will inevitably see a greater integration of national transport policies in Europe both as the Single market and subsequent developments in the EC lead to closer ties between the economies of Western Europe. Superimposed on this will be the gradual tying in oft the Eastern and Central European transport system. How successful the latter will be depends as much on the ability of management in the East as on the resources thrown at the problem by institutions such as the EBRD and others. There is the danger in this that the opportunities now available to up-grade the transport system to the needs of the next century could be lost if the process is captured by planners excessively concerned with centralisation and protection of specific modes.

Coupled with these moves towards coordination which must of themselves involve some physical expansion of the system are likely to be increased constraints on the way the system is used. Efficiency objectives, it is becoming increasingly realised, require appropriate prices and appropriate regulations. At present these are generally lacking but pressures are building for some, if not in the short term, complete rectification of this situaiton. Similarly, links between expenditure and infrastructure use are also only indirect for some modes at present but again, and stimulated by the need to generate additional revenues if transport infrastructure is to be provide efficiently, there are pressures for reform. One can anticipate, for these reasons, changes in the way fiscal arrangements are conducted with regard to European transport.

At the more micro-level, the needs of industry coupled with both changes in the general regulatory environmental and technological developments indicate that the next decade may well see rather more use of intermodal transport in its widest sense. In the past good arguments for more integrated transport have been advanced but the market forces were never allowed to fully operate and pressures came mainly from institutional sources. This has now changed.

REFERENCES

Adamson, M., Jones, W. and Pratt, R. (1991) "Competition issues in privatisation: lessons for the railways", in Banister, D. and Button, K.J. eds, *Transport in a Free Market Economy,* London: Macmillan.

Alvstam, C.G., Berman, Z. and Dawson, A.H. (1985), "The transport sector in Polish economic planning, and Polish-Soviet traffic capacity problems", in Ambler, J., Shaw, J.B. and Symons, L. eds, *Soviet and East European Transport Problems,* London: Croom Helm.

Ambler, J., Hunter, H. and Westwood, J.N. (1985), "Soviet railways – lethargy or crisis?", in, Ambler, J., Shaw, J.B. and Symons, L. eds, *Soviet and East European Transport Problems,* London: Croom Helm.

Argyris, N. (1989), "The EC rules on competition and the air transport sector", *Common Market Law Review,* 26: 5-32.

Association of European Airlines (1987), *Capacity of Aviation Systems in Europe. Scenario on Airport Congestion,* Brussels: AEA.

Barde, J-Ph. (1991), *Guidelines for the Application of Economic Instruments in Environmental Policy,* Background Paper 1, Environmental Committee Meeting at Ministerial Level, Paris, OECD.

Barde, J-Ph. and Button, K.J. eds. (1990), *Transport Policy and the Environment: Six Case Studies,* London: Earthscan.

Button, K.J. (1984), *Road Haulage Licensing and EC Transport Policy,* Aldershot: Gower.

Button, K.J. (1990), "Infrastructure plans for Europe", in J. Gidlund and G. Tornqvist, eds. *European Networks*. Umea: CERUM.

Button, K.J. (1991a), "The development of east-west European transport in the 1990s", in *Proceedings of the 26th Annual Conference of the Canadian Transportation Research Forum*, Quebec.

Button, K.J. (1991b), "The liberalisation of transport services" in D. Swann, *1992 and Beyond*, London. Routledge.

Button, K.J. (1991c), *Market and Intervention Failures in Transport Policy*, Paris: OECD.

Button, K.J. (1991d), "Privatisation in the transport sector: some of the key issues", *Economisch en Sociaal Tijdschrift* (forthcoming).

Button, K.J. (1991e), "Eelectronic road pricing: experience and prospects", *Proceedings of an International Conference on Economy and Environment in the 1990s*, University of Neuchatel.

Button, K.J. and Gillingwater, D. (1986), *Future Transport Policy*, London: Routeledge.

Button, K.J. and Ngoe, N. (1991), *Vehicle Ownership and Use Forecasting in Low Income Countries*, Crowthorne: Transport and Road Research Laboratory Contractor's Report.

Button, K.J. "High-technology companies: an examination of their transport needs", *Progress in Planning*, 29: 79-146.

Commission of the European Communities (1985), *Completing the Internal Market*, Brussels: COM (85) 310 Final.

Commission of the European Communities (1986a), *Medium-term Transport Infrastructure Programme*, Brussels: COM (86) 340 Final.

Commission of the European Communities (1986b), Report: *Towards a European High-speed Rail Network*, Brussels: COM (86) 341.

Commission of the European Communities (1990), *A New Structure for Community Railways*, Brussels: COM (89) 564.

Department of Transport (1989), *National Traffic Forecasts (Great Britain) 1989*, London: HMSO.

Drucker, P. (1990), "The emerging theory of manufacturing", *Harvard Business Review*, 68: 97-102.

European Conference of Ministers of Transport (1990), *Private and Public Investment in Transport*, Paris: Roundtable 81, ECMT.

Gent, H. van and Nijkamp, P. (1991), "Devolution of transport policy in Europe", in Button, K.J. and Pitfield, D. *Transport Deregulation: An International Movement*, London: Macmillan.

Group Transport 2000 Plus (1991), *Transport in a Fast Changing*

Europe, (No place of publication).

Gwilliam, K.M. and Allport, R.J. (1982), "A medium term transport research strategy for the EEC: Part 1", *Transport Reviews*, 2: 305-316.

House of Lords Select Committee on the European Communities (1990), *A New Structure for Community Railways*, HL Paper 11, London: HMSO.

Jones, P.M. and Hervik, A. (1991), "Restraining car traffic in European cities: an emerging role for road pricing", *Transportation Research* (fortcoming).

Kahn, A.E. (1988), "Surprises of airline deregulation", *American Economic Review, Papers and Proceedings*, 78: 316-322.

Laikin, R.E., Chadwick, M.J. and Cooke, J.G. (1987), "Energy-based emission inventories for modelling cost-effective SO_2 and NO_x abatement strategies in Europe", Paris: Paper presented at the International Workshop on Methodologies for Air Pollution Emission Inventories.

Lehmacher, H. (1990), "The Coordinating of the UN/ECE: Projects and Programmes", in Proceedings of a seminar on, *Developing a European Transport Infrastructure Network: The Case of Inland Transport*, Brussels: Centre for European Policy Studies.

Leijer, H.F.W.J. de and Tanja, P.T. (1989), *Gecombineerd Vervoer in Technologisch en Organisatorisch Perspectief: Potenties, Knelpunten en Oplossingsrichtingen, Notitie ten behove van de Sectorstudie Transport en Logistiek van het WRR-project Technologie, Overheid en Samenleving, Concept Eindrapportage*, INTRO-TNO, 89/LOG/476, Delft Technical University.

Linster, M. (1990), "Background facts and Figures", in European Conference of Ministers of Transport, *Transport Policy and the Environment*, Paris: ECMT.

Maggi, R. (1989), "Towards an economic theory of barriers to communication", *Papers of the Regional Science Association*, 66: 131-141.

Nijkamp, P. and Salomon, I. (1989), "Future spatial impacts of telecommunication", *Transportation Planning and Technology*, 13: 275-297.

Organisation for Economic Cooperation and Development (1987), *Toll Financing and Private Sector Involvement in Road Infrastructure Development*, Paris: OECD.

Organisation for Economic Cooperation and Development (1990), *Competition Policy and the Deregulation of Road Transport*, Paris: OECD.

Pickrell, D. (1990), "The regulation and deregulation of US

airlines", in Button, K.J. ed. *Airline Deregulation; International Experiences*, London: David Fulton..
Roundtable of European Industrialists (1990) *Missing Networks in Europe*, Paris: REI.
Ryan, T.A. (1989), "Air transport in Europe", 1989 *Louvain Lecture*, Louvain: Irish College.
Salomon, I. (1986), "Telecommunications and travel relationships: a review", *Transportation Research*, 22A: 247-255.
Schneider, L.M. (1985), "New era in transportation strategy", *Harvard Business Review*, 63: 118-126.
Seidenfus, H.St. (1991), "Prospects for railways", in European Conference of Ministers of Transport, *Prospects for East-West European Transport*, Paris: ECMT.
Timar, A. (1991), "Prospects for passenger transport", in European Conference of Ministers of Transport, *Prospects for East-West European Transport*, Paris: ECMT.
Transnet (1990), *Energy, Transport and the Environment*, London: Transnet.
Waele, A De (1991), "Consequences of closer East-West relations", in European Conference of Ministers of Transport, *Prospects for East-West European Transport*, Paris: ECMT.
Whitelegg, C. (1988), *Transport Policy in the EEC*, London: Routeledge.

TAMING THE WILD ROADWAYS: TEAMWORK, COMMUNICATION, AND SYNCHRONIZATION ACROSS SPACE

Richard J. Schonberger
President, Schonberger & Associates, Inc.
and
University of Washington
U.S.A.

How does the mind picture typical traffic upon the world's roadways? Perhaps we think of the apparently random wanderings of ants, except that on our roadways the aimlessness is bounded by the roadsides. Citizen drivers populating roadways include free spirits, aimlessly driving their cares away, as well as lethal weapons, taking out their frustrations recklessly.

May we exclude commercial haulers – whose mission is to move life's essentials and luxuries from one definite place to another – from our ant image? Not at all. The world's freight haulers operate within fluid time boundaries. Being late or early, going to the wrong place, getting stuck in traffic, breaking down, crashing, or otherwise behaving unpredictably is common, not exceptional in the over-the-road freight trade.

Our long-standing tolerance for wild roadways, free spirits, and general unpredictability, however, is finally wearing thin, at least as regards commercial traffic. A number of forces, mostly rather new, are acting in concert to drive today's over-the-road carriers into partnerships with shippers, customers, rail-sea-air freight movers, and electronic communications networks in the cause of quick, low-cost commerce – with full predictability and no surprises.

The primary forces, listed below, include several that are becoming competitively necessary, as well as a few supportive external influences.

Competitive Forces

- Knowledge-based business renaissance
- Just in time
- Minimal sourcing
- Certified suppliers and carriers
- Regularized scheduling and shipping patterns
- Total preventive maintenance
- Chain-of-customers teamwork
- Total quality and continuous improvement

External Forces

- Integrated shipping companies and alliances
- Multiple new technologies
- Enlarged ("borderless") trade zones
- Deregulation
- Global competitiveness

My intent is not to review these items one by one (a wealth of published information already exists on each). Rather it is to consider main issues as wholes and, in particular, how the world's roadways – and related organizations – are affected.

KNOWLEDGE-DRIVEN COMPETITIVE ENVIRONMENT

The new, elevated competitive environment has its roots in manufacturing. Business publications have named it a manufacturing renaissance, one that calls for continuous improvement in the eyes of the customer. In turn, the sweeping changes occurring in manufacturing have embroiled the road-based freight industry in its own renaissance. Additionally, final shipping points – the wholesale and retail trades – are becoming caught up in this period of dramatic change and improved performance. Lastly, the time is ripe for some of the not-for-profit governing bodies, such as roads bureaus, to have their own renaissance – in the cause of improved service to their public and commercial constituencies, as well as improved work lives for all.

In each sector – producers, haulers, final sales points, and governing bodies – there tend to be phases of change: First is denial that there is anything fundamentally wrong with "my firm". While denial is still mostly intact, the firm enters the phase in which it grudgingly makes changes demanded by key customers or providers. Next comes the switchover phase – from negative to positive; this is when a critical mass of opinion leaders in the organization become convinced that a wholesale shift – toward continuous improvement – is possible, necessary, and desirable. Maturity – continuous customer-centered improvement as a way of life – is the final phase.

Each phase is knowledge driven – led by a flood of well-attended seminars and best-selling books on excellence, total quality management, world-class performance, and related topics, backed up by a thriving education and consulting industry. In the maturity phase, the firm has its own prosperous internal knowledge generator and disseminator:

- Extensive, incessant competitive benchmarking
- Constant, close contact with parties along its customer chains and supplier chains
- Organized, frequent sharing of information among all employees
- All employees involved to some extent in the collection and sharing of all this lore.

MANUFACTURING RENAISSANCE

Let us briefly note some of the competitive forces in manufacturing that are ensnaring the freight carriers, as well as some of the leading final sellers. The hot fire of competition from Japan had pushed numerous Western manufacturers to the brink. Finally, at about the beginning of 1980s decade, manufacturers around the globe discovered that culture had little or nothing to do with Japan's success. It was teachable concepts and techniques.

One set of approaches that has been affecting road traffic practices is just in time (JIT). JIT calls for producing and moving parts just in time for use at each stage in the long chain of customers leading from the mines and forests to final users. The chain may include hundreds or thousands of moves of materials (and information) within the various manufacturing plants along the chain, as well as dozens of transports from site to site.

Milliken and Company, a privately-held multinational textile and chemical maker (and winner of a Baldrige U.S. national award for quality), announced in 1987 its plan to drive response time –

from textile ordering all the way to receipt in the retail store – downward from 16 weeks to three weeks. That goal has been partially achieved. By partially, I mean through a few of the key flow paths that start with an order for a Milliken fabric. For example, such a flow path might include a Milliken plant spinning the thread, another weaving the fabric, another dyeing it; an apparel company that cuts, sews, and packages the fabric into shirts; and, lastly, receipt of the packages of shirts by a department store. Included in the three weeks are transports from plant to plant and to the final seller, perhaps including long highway hauls.

Best products and best companies are first candidates for this kind of quick-response treatment. Best products would include the dominant fabric weaves and colors, and the dominant shirt styles and sizes. Best companies would include the Milliken plants that have the quickest changeovers, most versatile and well-trained employees, best quality, and most dependable equipment; the same applies to the apparel maker.

WORLD-CLASS FREIGHT HAULERS

The best freight carriers are a new breed. Capitalizing on trucking deregulation in the U.S. in 1980, plus new opportunities to be awarded a long-term, semi-exclusive contract for repetitive hauling of primary products, trucking companies like J.B. Hunt in the U.S. have become giants in just a few years. Hunt keeps its trucks washed. Hunt drivers wear uniforms and are clean-shaven – and stand out at truck stops amid their long-haired, leather-vested, cowboy-booted brethren (Machalaba, 1991).

Other J.B. Hunts surely will spring up, especially as trade pacts and intercountry standardization create vast new mass markets. Mass markets spawn specialized plants producing enormous volumes that must be trucked by dedicated carriers through the chains of customers leading to final sellers. Those mastering streamlined simplicity, excellence, and continuous improvement will thrive.

World-class freight haulers must pick up and deliver on time, which sometimes means to the hour, and must not damage goods. Increasingly, trucks are equipped with on-board computers and satellite navigation systems, as well as cellular phones. Meticulous records are kept on every tractor, so that failure rates become the basis for effective predictive maintenance. In following these practices, the U.S.-based giant, United Parcel Service, keeps its thousands of long- and short-haul vehicles out of the breakdown mode and in operation for decades before retirement - much as airline companies do.

DIVERGING TRENDS: ELECTRONIC AND VISUAL COMMUNICATION

Retailers, too, are changing their ways. A fast-growing trend is for laser-scanned final sales data to be electronically fed back through the supply chain. This permits scheduling plant production and interplant deliveries to closely match real sales patterns - an ultimate sort of synchronized JIT scheduling system. This accelerates the trend toward elimination of intermediate warehouses, which, in turn, cuts the amount of trucking that must be done. On the other hand, JIT's demands for frequent, small-lot deliveries causes a shift of some freight handling from rail to roadways.

Ironically, while electronic data interchange (EDI) helps link final sales points to the supply chain and the freight haulers, visual communication is at the same time replacing much of what conventionally required written transactions. The visual authorization system, pioneered by the Toyota family of companies in Japan, is called "kanban" - a Japanese word that has been internationalized. Among the several forms of kanban is the labeled, recirculating container.

CalComp, a dominant maker of graphics terminals, has several of its Southern California suppliers on kanban. A sheet metal supplier delivers a few specially padded wheeled carts of metal "skins" to CalComp's final assembly area daily, and takes away now-empty carts that were delivered the day before. The carts are labeled (the "kanban" identifier) and contain an pre-set, never-varying quantity, so that no receiving count is done.

At A. Ahlstrom Corporation's pump factory in Mantta, Finland, many suppliers of bearings, gaskets, and other small parts collect kanban cards, rather than kanban containers, when they appear at Mantta to make a delivery. Ahlstrom machinists remove cards from parts cartons as the parts are used up. The supplier, or perhaps the truck driver, provides just what the kanban cards indicate - no transactions necessary. The role of EDI may be limited to handling monthly invoicing and funds transfer.

Perhaps the burdensome paperwork that hampers freight

hauling, especially across borders, will be simplified in a similar manner – just collect pre-printed cards and pay fees monthly – especially for trucks that are on regular delivery routes, called "milk runs". The regularizing of trucking, another strong pattern associated with the manufacturing renaissance, bears further comment.

REGULARIZED TRUCKING – AND ITS EFFECTS ON HUMAN INTERACTIONS

A growing number of manufacturers receive incoming materials in reusable containers at multiple docks around the perimeter of the plant. Materials come directly to points of use with no inspection. Sometimes the buying company collaborates with multiple suppliers, plus freight carriers, to establish regular milk runs: A truck picks up from multiple sites and may deliver to multiple plants and docks on a single trip – which repeats at regular intervals, sometimes more than once a day.

Ideally, the route driver gets to know the people at the sources and destinations – and develops an understanding of needs. As this pattern evolves, the driver may be invited to company picnics and even to serve on improvement teams – jointly with people at both ends. Chemical Leaman Tank Lines, a Pennsylvania-based hauler, sends action teams of its employees to work with clients and sets up employee exchanges eith customers.

Leaman also has initiated quality improvement teams involving terminal management, dispatchers, and hourly employees (HR Reporter, 1991). This may not sound like much of an innovation, but it is. Managers at a large grocery distributor on the West Coast of the U.S. decided that improvement teams were necessary but followed the more typical approach: one improvement team composed of drivers, another of dispatchers, another of cold storage people, and so forth. I interviewed one of the drivers, who could not conceal his disgust and frustration over his team's proceedings. The drivers' dominant problems were at the interfaces with those in dispatching, loading, paperwork, maintenance, customer relations, and the other departments. Since none of those people were on his improvement team, the driver said his team's meetings degenerated into gripe sessions. What the grocery distributor created is not worthy of term, team; they are more like gangs – who waste much of their time blaming other gangs, having been denied the chance to team up with people with whom they have a work flow relationship.

The Leaman pattern, teams formed of people who have a chain-of-customers (work-flow) relationship, became popular in advanced manufacturing companies a few years ago. In manufacturing, the new teams are often permanent, involving elimination of the functional organization of shops and relocation of machines and people into product-family-focused work cells. Some manufacturers have also broken up staff support departments and moved engineers' desks to the production floor where they may become members of the focused production team. The teams sometimes extend to include product design people, buyers, and supplier representatives, some of whom are given desks in the buying company's plant during the product development phase. One medical products company is trying to get agents of the U.S. Federal Drug Administration assigned to its extended product development teams. Similarly, leading trucking companies may want to include regulatory agents from roads bureaus on their improvement teams.

TOTAL QUALITY MANAGEMENT IN THE NON-PROFIT SECTORS

While industrial, freight, and sales companies press their regulatory counterparts into teaming up for improvement, some public and non-profit managers are discovering total quality management (TQM) on their own. As public service managers learn about TQM, they may become enthused over the chance to transform their working lives and careers – and become advocates for initiating continuous improvement teams with their clientele.

In the U.S., a medium-sized city, Madison, Wisconsin, has achieved a measure of renown for the way it has embraced TQM in some of its service areas. For example, Madison's kickoff "project" was to straighten out the mess in the city garage, which was in charge of repair and maintenance of 900 municipal vehicles. One result was reduction of average turnaround time from nine days to 2.5 days. In the process, the president of the associated labor union became an articulate spokesman for quality improvement (Box, et al., 1991).

A FEW STRONG PARTNERS

Total quality and just-in-time requirements have induced leading manufacturers to greatly reduce the numbers of suppliers they deal with. Frequent close contact with suppliers is seen as highly advantageous - in product development stages as well as in ongoing operations. Last decade's frenzied global searches for high quality, low cost suppliers were temporal; today's world-class view is that the local supplier base, so badly neglected and mistreated in the past, must be invested in, rebuilt, and partnered-up with.

Since each of a reduced number of nearby suppliers has far more business volume with a customer than before, costs are driven down and strong interdependencies are created. This subverts the old adversarial relationship and paves the way for working together for mutual, long-term gain. Leading manufacturers have been extending the concept beyond their suppliers to their freight carriers. 3M Corporation used to deal with some one thousand less-than-truckload (LTL) and 240 truckload (TL) carriers. They reduced the numbers to 31 core haulers moving 87 percent of 3M's LTL shipments, and just seven handling over 91 percent of its TL freight (Moskal, 1989).

SUMMARY

In summary of the above, my crystal ball conjures up this vision of how the world's roadways and roadway-affiliated organizations are to be affected by the "excellence" movement:

- High annual volumes of parts are trucked frequently in small quantities over fairly short distances. Each buyer site has a small number of excellent suppliers-partners; each supplier site has a small number of excellent customer-partners.
- End products are trucked, rail-trained, or sea-land transported on regularized runs from country to country across vastly enlarged mass markets.
- Along with regularity comes administrative simplicity, which is fostered by a combination of electronic communication, visual kanban, and improvement teams whose membership transcends organizational boundaries.

- Operations will be uniformly simple, quick, capable, and low-cost; irregularity, with attendant high overhead costs and complications, will be confined to where it fits – in new product development, opening new marketing channels, and so forth.
- The weak – those who are delay-prone, red-tape-bound, incapable of mishap-free operations, and uninvolved in continuous improvement – will disappear, in favor of the strong (their world-class opposites).

REFERENCES

(All U.S.-based periodicals)

Box, G.E.P., Joiner, L.W., Rohan, S., and Sensenbrenner, F.J., "Quality in the Community: One City's Experience", *Quality Progress*, May 1991, pp. 57-63.

Machalaba, D., "J.B. Hunt Reinvented Trucking, but Now Has Load of Problems", *The Wall Street Journal*, May 9, 1991.

Moskal, B., "Enough Premium Carriers in the '90s? Don't Bet on It", *Industry Week*, August 21, 1989, pp. 31-34.

"Quality: Partnership With Customers", *HR Reporter*, February 1991, p. 6.

PART II:

REGIONAL PERSPECTIVES

THE CENTRAL EUROPEAN PERSPECTIVE

Heinz Sandhäger
Director General, Federal Ministry of Transport
Germany

PRESENT SITUATION

During the eighties, the European transport market was an enormous growth market. This is true for goods and for passenger transport.

All growth forecasts for the central regions of Europe have been overtaken by actual developments. This does not only apply to long-term forecasts over ten to fifteen years, but also to short-term forecasts. A particularly striking example is the 1985 German Traffic Infrastructure Plan. It was meant to cover of the 15 years period until the year 2000. However, it forecast for road traffic was fulfilled after only one year, i.e. at the end of 1986, instead of in the year 2000.

In addition, traffic growth reflects another particular characteristic. It concentrats on two modes of transport, road traffic and air traffic. The transport volume of the waterways is stagnant. The railway system was and still is the loser in the growth process with a present market share – related to the long distance transport volume – of 30 percent. In border-crossing goods transport within the EC, the railway system's market share reaches to only 19 percent.

Thus we experience the strong dynamics of the growth of traffic on the one hand, but, on the other hand the distribution of this growth is strongly biased towards to road and air transport. It is these two factors that confront us with tremendous problems of bottlenecks in the centre of Europe.

FORECASTS OF FUTURE DEVELOPMENTS

The Federal Republic of Germany intends – after the unification and the end of the division of Europe – to prepare a first Traffic Infrastructure Plan for the united Germany. The planning period extend to the year 2005, with a reserve period up to the year 2010.

In formulating this plan we face an absolutely new situation. In the past, traffic streams in Europe mainly ran in a north-south direction, as Figure 1 impressively shows. For more than 45 years the main traffic arteries were therefore built and developed in this manner. And yet despite this we did not succeed in eliminating all bottlenecks in the race against the growth of traffic.

Figure 1. *Traffic flows in Europe.*

The unification of Germany and the end of the divided of Europe will generate new traffic growth from the west to the east and east to the west. Today's infrastructure is not at all prepared for this development.

This applies to the infrastructure of the central regions of Western Europe, but even more, to the infrastructure of Eastern Germany and its neighboring countries in Eastern Europe. Together they will generate considerable transport growth without being able to fulfill major transit functions.

The Traffic Infrastruture Plan for the united Germany starts out from forecasts for the year 2010, as shown in Table 1. If we consider the goods sector, the long-distance goods transport by road will increase by 95 percent, 55 percent by rail and 84 percent by inland waterways. This means that the road network will no longer experience the same growth rates as during the 1980s.

Table 1. *Forecast for the year 2010. Goods transport and passenger transport in Germany*

Goods transport	
Road	+ 95 %
Rail	+ 55 %
Inland waterways	+ 84 %
Passenger transport	
Motorcar	+ 30 %
Rail	+ 41 %
Air	+ 151 %
Public short-distance transport	+ 27 %

In passenger transport, private transport will increase by 30 percent. Transport by rail will increase by 41 percent while long-distance transport will increase by 58 percent. Air transport will increase by 151 percent and public short-distance transport by 27 percent.

The forecasts refer to the base year 1988, and take the growth impulses originating in Europe into account.

STRATEGY FOR A FUTURE TRANSPORT POLICY IN THE CENTRE OF EUROPE

During the nineties (more so than during the eighties) and beyond the turn of the millenium, transport policy will be confronted with the question of how traffic growth can be coped with in central Europe facing the new perspective for the whole of Europe. It is the aim of the transport policy of the Federal Republic of Germany, as an integral part of the Single European Market, to prevent a complete traffic collapse at all costs. It can not be allowed that traffic bottlenecks will slow down the national and European economic development.

Thus we pursue six principal strategies in order to deal with this first-priority challenge.

The Strategy of Avoidance

Vehicles employed in European goods transport today, are doing up to 35 percent of their runs empty. Over short distances this share increases to as much as 60 percent. Here we have a large potential for logistics and telematics to assist in maximizing the use existing capacities as far as possible.

If we do not use these potentials, society will be less and less inclined to consent to new roads and other traffic routes.

In passenger transport there is also great scope for strategies of avoidance. In urban agglomerations the share of traffic seeking parking facilities amounts to as much as 40 percent. Intelligent information systems can avoid this traffic bottlenecks to quite a considerable extent.

These are only a few examples. Telematics can also replace traffic bottlenecks. In the longer term, strategies of avoidance must become an integral part of regional structural policy, regional planning policy and urban development policy.

The Strategy of Connecting Transport Modes

The systematic interconnection of various modes of transport offers the only real chance of overcoming one-sided growth in conformity with market mechanisms. For this reason traffic legislation, technical standards as well as public transport investments must promote integrated systems.

Integrated solutions exploit the characteristic advantages of rail, road, waterway and air transport.

One exampel of integration in goods transport is combined transport.

Combined transport provides a chance for the railway system to compete with other modes of transport. More importantly, it provides the chance for gain higher market shares.

Combined transport connections are significant between urban agglomerations, where high-quality goods account for about 70 percent of goods transport. In Germany, 1.9 million journeys by lorry per year are currently being replaced by combined transport. For comparison, 1.5 million lorries transit Germany each year.

All empirical studies show that the potential for integrated solutions in combined transport is at least three to four times as great as the existing volume of combined transport in the centre of Europe. The networks of the waterways, of ferry traffic and of European coastal shipping must be integrated in goods transport. There is considerable potentials in this area.

The generally recognized weak points of the integrated transport system are its connections. The main share of future investments must be dedicated to their improvement. The eastern European countries have also recognized this necessity. In eastern Germany we hope to make full use of the new starting point in order to develop an efficient network of terminals right from the beginning.

The integrated transport system requires universal transport equipment offering equal advantages to all modes of transport, with dimensions which will not change in the long run and which must be the same worldwide.

Regarding passenger transport, the development of national and European high-speed networks must pave the way for passenger transport by rail to increase its currently low share of 10 percent. The objective of transport policy must be to displace the demand for short-distance air transport within the central region of Europe to long-distance rail transport.

It is our intention to connect the most important European centres by a European high-speed rail network.

A draft plan of a European high-speed railway network was presented to the EC Council at the end of 1990 and is now being further investigated. What is apparent here, is that Europe is beginning to recognize the future challenge it will have to face with regard to mobility. The answer to mass mobility can only be an efficient means of large-scale passenger transport such as that provided by the railway system.

The Investment Strategy

It is becoming more and more difficult to obtain general consent for the construction of new traffic routes in the Federal Republic of Germany. For example, 40 percent of the new 430 km long railway line between Hanover and Würzburg runs through tunnels, as it would not have been possible to realize this project otherwise. The protection of the environment has become an integral component of traffic planning. It has to be taken into consideration right from the start. Nowadays, the Federal Minister of Transport is the greatest landscape gardener in Germany. Each year he has 12 million trees and bushes planted in order to make the construction of traffic routes at all possible. These measures require a lot of money.

In the Federal Republic of Germany the extension of the traffic infrastructure did not kept the same pace as the growth of traffic.

In the eighties

- the gross national product rose by 50 percent,
- the public budget increased by 35 percent,
- but the transport budget did not even increase in nominal terms and investment in the transport sector, in real terms, even decreased by 30 percent.

The new German Traffic Infrastructure Plan for the united Germany up to the year 2010 will reverse this trend. It will increase investment in traffic infrastructure and mainly concentrate on

- the improvement of railway infrastructure,
- the improvement of the connections of integrated transport systems in goods and passenger transport.

The most important elements of the new German Traffic Infrastructure Plan for the united Germany are the so-called "Traffic Projects for German Unity". They are designed for the re-establishment of efficient east-west arteries in an overall European framework. Seventeen large-scale projects have been envisaged to serve as links in all direction and to Scandinavia and eastern Europe as well.

Figure 2 presents the nine projects related to the railway sector.

- the connection along the Baltic coast between Lübeck and Stralsund
- the connections from Hanover to Berlin in the course of the high-speed relation Paris-Warsaw
- the connections to Leipzig and from Leipzig to Berlin
- and the Göttingen-Halle-Leipzig-Dresden line.

Figure 3 shows the most important highway projects, which are:

- the connection between Lübeck and the western border of Poland
- the connection between Kassel, Erfurt and Dresden
- the connection between Hanover and Berlin
- the improvement of the autobahn between Nuremberg and Berlin.

These projects are to be realized as soon as possible by means of a special law.

We will begin planning these projects already before the Traffic Infrastructure Plan for the united Germany has been adopted. We cannot afford planning periods of between 10 and 15 years, previously the practice in the western part of Germany.

Only with the aid of these laws we succeed in obtaining noticeable improvements within the next two to five years. The planning times are to be shortened by constitutional means, without infringing upon the citizens' rights of objection.

The Strategy of the Market

The decision to work towards the European Single Market also entails the requirement that the transport sector must be subjected to the rules of largely deregulated markets. But this can only work if it is possible to create fair conditions of competition throughout Europe. This task necessitates

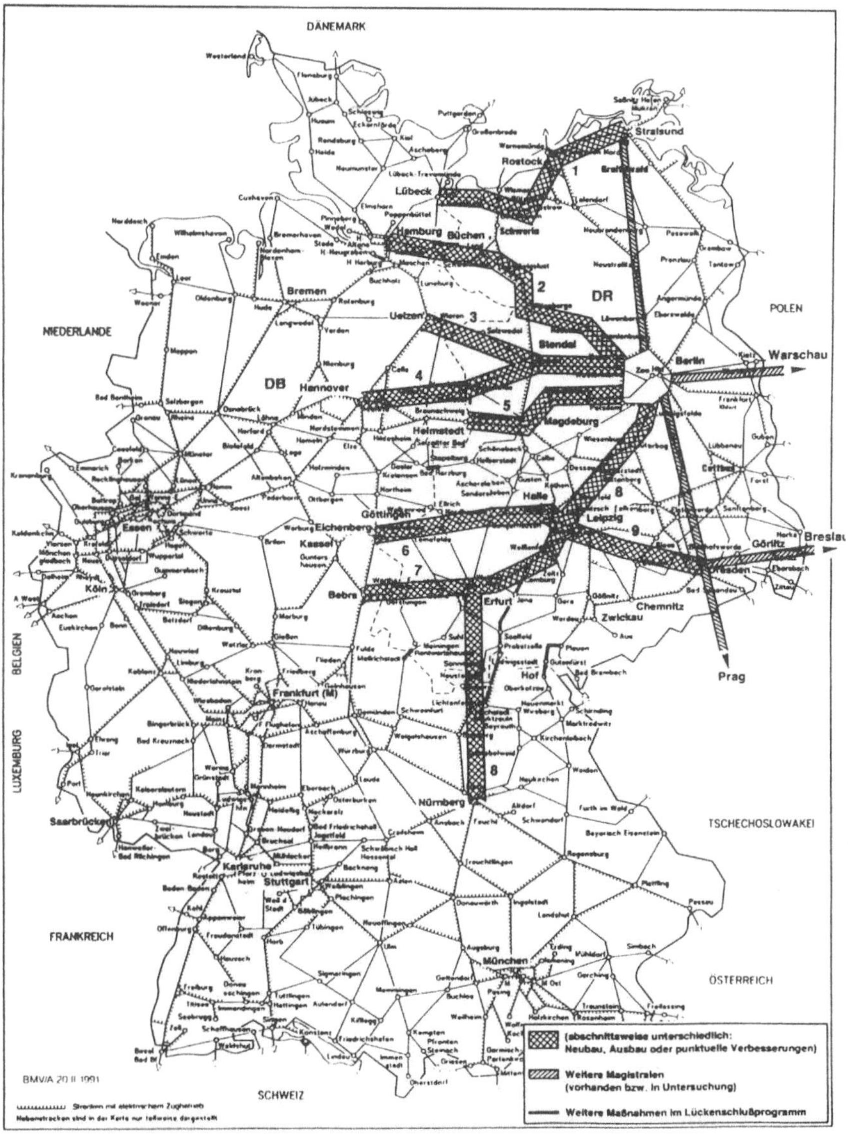

Figure 2. *Projects of the railway sector. From the German Traffic Infrastructure Plan for the united Germany.*

- a harmonized system of transport-specific charges and taxes,
- uniform quality, environment and safety standards,
- uniform provisions for the transport of dangerous goods,
- uniform controls,
- and the elimination of undue subsidies.

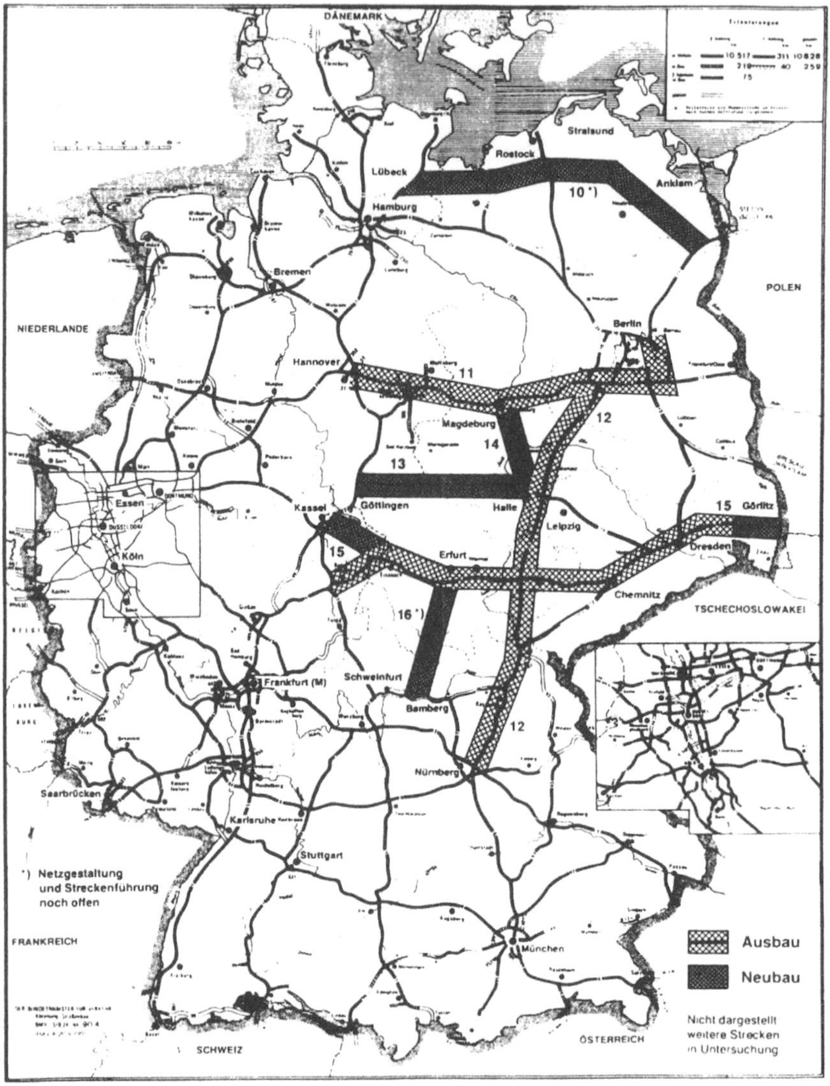

Figure 3. *Highway projects. From the German Traffic Infrastructure Plan for the united Germany.*

For the time being, these requirements are still far from being fulfilled. This can lead to distortions in the transport markets, especially in the central regions of Europe. Furthermore, there are sectors where the price has hitherto not acted as an instrument of the market, namely as a signal of shortage. It is, however, necessary that market prices apply to all sectors of transport.

This definitely means that in respect to

- the use of the traffic infrastructure provided by the state,
- the environmental strain caused by traffic,
- the scarcity and environmental compatibility of energy prices must also fulfil their market function.

As regards the state-owned traffic infrastructur, in many cases the charges paid by the user do not cover the costs and certainly do not reflect the scarcity. Privately-funded solutions are much better at taking the rules of the market into account and, at the same time, offer new possibilities of financing.

European transport policy must therefore establish pricing systems which are in line with market requirements for all infrastructures of the transport sector.

In order to cope with the environmental strain caused by traffic the bases of assessment for public charging systems in the transport sector must be increasingly oriented towards the degree of the environmental strain. European solutions are also necessary in this respect.

In order to achieve the economical handling of energy resources the market conditions for alternative fuels and alternative propulsion methods must be essentially improved through a well-balanced system of incentives and charges.

The Environmental Strategy

We will only be able to cope with the growth of traffic in the central regions of Europe if we always apply – both to the infrastructure and to the means of transport – the latest state of technology with the highest environmental standards.

In connection to this I should like to evolve two theses, the first describing the risk and the second the opportunity for the future.

The first thesis is:

A densely populated and industrialized Europe which continues to apply average technical environmental standards will produce tomorrow's home-made mobility crisis.

The second thesis is:

Any state which consistently invests in environmentally friendly technologies in the transport sector today will take the competitive lead in tomorrow.

I should like to refer to the fact that Germany pursues the

ambitious objective of reducing the national carbon dioxide emissions by 25 percent by the year 2005 as compared to the year 1987. This objective can only be reached by rigorously applying the above-mentioned environmental strategy.

Traffic safety is an integral component of a consistent environmental policy in the transport sector. The implementation of a policy with a view to increase traffic safety is therefore more urgently required than ever before. The drastic rise in the number of accidents in eastern Germany presents a special challenge to our country. Appropriate measures must be taken with regard to traffic routes, transport means, and the behaviour of transport users.

The Strategies of the Future

These strategies take into account the fact that we must immediately begin to deal with certain problem areas ever though we know that solutions can only be realized in the medium or longer term. I am thinking of

- the use of renewable energies in the transport sector
- the recycling of cars
- and new forms of use of the individual motor car.

The level of mineral oil prices during the last decade has rather weakened initiatives towards the search for other, and – if possible – renewable, energies. The "mountain of used cars" is also becoming higher and higher in Europe. This forces us to take rigorous action with regard to car recycling. And finally:

In the Federal Republic of Germany, 35 milion motor cars are moved for 40 minutes on an average each day. This means that for the remainder of the day these vehicles form the greatest store with the highest land requirement. It is can already be questioned whether this is the most intelligent form of car use, but at the same time we know that new solutions are not readily at hand.

However, if we want to find solutions for the near future, we must begin to look for the right answers today.

THE EASTERN EUROPEAN PERSPECTIVE

Igor P. Kiselev
Chairman of the Permanent Transport Complex Committee of Leningrad
USSR

INTRODUCTION

In Eastern European countries of today, we see on the one hand the tendency towards better exchanges of information, growing volumes of goods transported and an increased volume of passenger traffic. On the other hand, we have to admit that the present system of transport has completely exhausted itself. The shortage of goods and services that is characteristic of the USSR, is most evident when we speak about transport production.

In this situation the analyses of current conditions in the field of transport and communications in the North-west part of the USSR or rather the Leningrad area, are believed to be very important for better understanding of transport problems and their prospective solutions through out Eastern Europe.

Leningrad is the largest cultural, industrial and scientific center of the Union and second largest city. The population of the Leningrad area is 6.7 million (Leningrad City 5.1). Having 2.3 percent of the Union population, the Leningrad area produces 3 percent of the national income and more than 5 percent of its machines and other technical equipment. 13 percent of all scientific work and research in the Union is carried out here.

In addition, Leningrad is the Union's largest historical-architectural center. It is one of the few cities in the world built according to an architectural plan. In 1991 UNESCO added to the world heritage list the exceptional example of town planning in the City of St. Petersburg and its environments.

Leningrad is a great tourist center. 13-14 million people annually visit the city. However, the capacity of hotel accommodation is 30,000 but that is not enough. Due to the unfavourable economic, social and political situation, the amount of tourists staying in Leningrad has declined sharply in recent years. From 1.2 million foreign tourists 1988 to 0.5 in 1990.

LENINGRAD'S TRANSPORT FACILITIES - PROBLEMS AND PERSPECTIVES

General

The problems of city transportation are not included in this paper - however it is necessary to stress that the main focus in Leningrad is upon the public transport system. In 1985 the Leningrad Metro underground system, trams, trolley-buses and buses carried more than 3,200 million passengers.

The current capacity of the city transport system, the extent of rolling stock, the development of the roads and streets network and their quality do not correspond to the needs of the city.

The Leningrad complex transport junction is the largest and the oldest in the Union. St. Petersburg was the cradle of modern merchant shipping, railway, automobile and air transportation.

During the period 1965-85 the volume of passengers carried by all modes of transport in the Leningrad area increased. The number of passengers increased from 2,706 to 4,450 millions' per year. By the year 2005 this figure is expected to have reached the level of 6,000 millions.

RAILWAY TRANSPORT

In 1837 the first Russian railway was opened in St. Petersburg. By 1851 St. Petersburg and Moscow were connected by rail.

Today the October Railway, with its center in Leningrad, measures 10,500 km and provides transportation in the Moscow, Tver, Pskov,

Novgorod, Leningrad and Murmansk areas as well as the Carelian Republic.

Leningrad has five railway stations and ten main railway lines teminate here. The city is connected with all areas of the Union by these lines. It takes more than ten days to go to Siberia and the Far East by train. In addition, Leningrad is connected to Warsaw, Berlin, Hannover, Sofia, Prague and Helsinki by rail .

The railway was and still is the base of the transportation system in the Union. In order to increase the level of transport services in, and foreign connections to the North-Western part of the USSR, further improvements in the railway system are required.

In this respect two projects related to the railway system are noted: The reconstruction and development of the **Leningrad Railway Junction** and the improvement of **the main lines.**

The Leningrad Railway Junction

The major problem of this 'super' junction is the fact that the main lines end here. Passengers must be served and goods has to be disbributed to freight stations located right in the city itself. In addition, passenger and freight facilities are out of date and there are no or extremely limited possibilities for horizontal expansion of the terminals.

A new, large passenger station, the Ladozhsky Station, is now being constructed. This station will be the sixth in the city. A large marshalling yard located beyond the city limits is also being designed.

By reconstructing existing railway lines and introducing tunnels with a complete clearance under the city center, the long-term goal is to convert Leningrad from a line-end to a line-through junction. The tunnels will interconnect all of the northbound and southbound lines and thus speed up transit the transportations significantly.

The Main Lines

The projects of the Leningrad junctionprojects are interrelated with the construction of the Union's first high speed main line from Leningrad through Moscow to the South (Crimae, Caucasus Mountains) with a possible branch from Leningrad to Helsinki. The design work for this railway is almost complete. The examination of the main sectionbetween Leningrad and Moscow was carried out by the Academy of Sciences of the USSR in cooperation with French experts from the TGV.

The construction of this new Leningrad - Moscow separate double track line is planned to start during 1992. It is designed to carry electric trains with two-level passenger cars and container and refrigerator cars at a speed of 350 km/h. The train will manage the distance Leningrad and Moscow in 2.5 hours and Leningrad - Crimae in 8 hours.

The main technical problems have now been solved. Remaining is however financial problem.

Naturally, the construction of the highspeed line connection Leningrad, Moscow and the South also incorporates the construction of a highspeed line connecting Moscow to Western Europe via Brest (200 km East of Warsaw). At a certain stage during the work with the design of these two lines they were considered to be competive. In whatever manner it today looks like the Leningrad, Moscow and the South will be the first one to be constructed.

Interesting is also the fact that during the planning of these lines a study of another high speed line from Leningrad to Helsinki was under consideration. The distance from Helsinki to Moscow could be done in one night (6 - 7 hours).

SEA AND RIVER TRANSPORT

Peter I planned St. Petersburg as a "window to Europe", that is as a port on the border between the inland water ways of Russia and the worldwide sea trade system. Leningrad is still very important in this respect and we will find that Leningrad takes almost 1/3 of the trade in all of the Baltic ports on USSR territory.

The most important waterway of the "river-sea" system is the Volgo-Baltisky shipping route with guaranteed depth and width, which passes through Leningrad. The traffic includes container lines from Western Europe to Japan and from Great Britain to the Caspian Sea.

Unfortunately, the great potential of passenger-oriented transportation at sea and on the rivers is barely not used. Leningrad has good sea and river passenger terminals, but in 1985 only 220,000 passengers were carried by sea and 350,000 by river vessels, which is only 1.7 percent of the total long distance passenger travelling through the Leningrad junction.

Passenger ships at sea operate from Leningrad to Helsinki, Stockholm and other cities. River passenger ships operate to Astrakhan, Nizhny Novgorod, Rostov-on-Don etc.

The main problems of the sea and river transports are due to the fact that Leningrad's ports, docks and container facilities need to be modernized to a great extent. It is also necessary to renew the rolling stock serving this traffic.

Due to historical reasons the sea and river ports of Leningrad are located inside the city limits. This creates difficult problems in connecting the ports to the main rail and road systems of the city. The problems has intensified by the political situation in the Soviet Baltic Republics. For this reason, it is hardly possible to increase the goods flow for the USSR via the ports of Riga, Ventspils and Tallinn. In fact the goods of these ports is expected to pass through Leningrad instead.

At the moment the construction of a new Leningrad trade port in the outskirts of the city Luzhskaya located in the East of the Finnish Gulf is under consideration.

This would mean large economic investments and jobs for many construction workers, but also new social problems to solve.

Since the routes from the Baltic Sea and the Atlantic Ocean to the river Volga, the Caspian, the Azov and the Black Seas pass through the city center the Leningrad bridges present certain difficulties in the development of river-sea navigation. Ships can only pass on the river Neva at night when the bridges are raised.

This is the main reason to build tunnels under the Neva. The Metro, roads and rail that today occupy bridges would then use the tunnels, and the bridges could be lifted in day time as well as at night.

At present a new passenger river port located in the Western part of Bazil's Island is being studied. With a location downstream of all of Leningrad's bridges, this port will facilitate sea travel for tourists between Leningrad and ports overseas.

AIR TRANSPORT

Leningrad has two airports. Pulkovo for domestic and international flights and Rzhevka for domestic mail, freight, ambulance and fire fighting flights.

The Pulkovo which is one of the largest airports in the USSR, is a modern establishment. It has two runways and a manifold of navigational-technical equipment permitting all types of international air traffic, without exception.

The volume of domestic operations at the Leningrad airports has increased by a factor of 3 in the last 20 years. In the same period USSR as a whole increased its volume by a factor of 2.6.

Leningrad has nonstop flights to 170 cities in the USSR (in 1970 this number was 58) and to 21 large European cities. In 1990 the Leningrad - New York route was initiated.

The passenger lines are distributed as follows:

- almost 25 percent: Central and North West areas of the Union (Moscow, Murmansk, Archangelsk etc)
- 15 percent: The Southern directions.
- 10 percent: Central Asia, Siberia, and the Far East.
- 5 percent: International destinations. Most popular are Berlin, Prague, Warsaw, London and New York.
- 45 percent: Other local and long distant lines.

In 1965 2.2 million of passengers passed through the airports of Leningrad. In 1985 this figure reached 6.5 million.

It is impossible to increase the number of passengers at the Pulkovo Airport as the existing resources of the terminal facilities are exhausted.

At the moment plans for the construction of a new international airport connected to Pulkovo Airport are under being completed. This new section of the airport will include passenger terminals with shopping areas as well as facilities for flight catering services and hotels.

This project will be designed and fulfilled in colaboration with a number of foreign companies. The project will make it possible in the next five years to supplement the beautiful airport buildings with all the necessary components required for passenger service. This will greatly increase Leningrad's airport capacity, and is considered a step towards the creation of a municipal airport based on a city owned company.

In the more distant future it will be necessary to design and construct a new international airport at Tosno located some 70 km outside Leningrad, and to build modern high-speed means of communications connecting this airport to the city.

ROAD TRANSPORT

The situation of the road transport sector in the Leningrad area seems to be most dramatic and contradicory.

It is the most dynamic sector of transportation with reference to the increase in volumes of transported goods as well as in number of vehicles. Still it is in the same sector that the USSR has fallen behind in comparison to its West European and Scandinavian neighbors.

The difference between the technical equipment of vehicles, their comfort and their impact on the environment is growing rapidly. This is due to the breathtaking shortage of vehicles. A shortage which forcesthe public administration, trade and industry to buy any car, bus or truck that can be found.

Today, the Leningrad area badly needs several thousand buses, thousands of taxis and trucks. This shortage in the market gives birth to an avalanche process of using the existing fleet in a way that exceeds all possible limits in terms of operation.

The situation becomes even worse since all these vehicles are driven on very bad highways. The North-West area of the USSR, including the Leningrad area, has not one single road which meets international requirements. Even our best highways, such as the roads from Leningrad to Moscow, Vyborg, Tallinn and Kiev, do not reach the level of design and construction nor the levels of service stated in the present requirements for such roads.

In the fields of Road Traffic Management, Road Traffic Information and General Services to the road users, the USSR is in need of sharing the experience of others. Most urgent is to introduce international harmonization concerning Traffic Regulations and related services such as safety arrangements and information to drivers given along the road by signs and other devices. Safety improvements and better

Road Navigation Systems seem to be of great importance.

Referring to an expected rapid development in all sectors of the national and international road transport system, it seems most important to prepare the Road Administrations and related units about adequate principles of Traffic Management etc. of modern road transport. For this reason, it is urgent for the USSR to support an immediate process of international harmonization to make sure the tools and principles used to manage road traffic will correspond to the needs of the future. Preferably such a process should be taking place in advance of any major improvement of the road network and in close cooperation with other nations of the UN.

Training and information in Traffic Management and Safety is a must for all professionals. The coordination of Traffic Operations is a difficult task in itself which will be impossible to fulfill sufficiently if the staff is unqualified and poorly trained and equipped.

In total, 8 European highways run through the USSR: El8, E20, E30, E40, E50, E85, E93 and E95. The following concern Leningrad:

El8:	... - Stockholm - Helsinki - Leningrad.
E20:	... - Stockholm - Tallinn - Leningrad.
E95:	Leningrad - Moscow - Kharkov - Jalta (Crimea).

If we leave the main highways for the feeder roads in the countryside and the minor streets in Leningrad itself, we find ourselves upon what are called "impassable roads" in "Russia's"!

There is no shortage of good plans - excellent plans and projects for road modernization and road transport facilities have been worked out in recent years. One example is the design of the highway from the Finnish border via Vyborg to Leningrad and further down to Moscow. In addition, a high speed ring road around Leningrad has been designed. Parts of this road will be located on the dam in the Gulf of Finland and on Kotlin Island.

These plans have not yet been realized. There are objective and subjective reasons for that. We have no money, nor investors. In addition, there are no distinct standards or rules for the protection of the environment which has blocked the construction of a new asphalt-cement plant and the operation of the old ones.

In table 1 the investments for the modernization of the Leningrad area transportation facilites are presented.

Table 1. *Approximate investments required for the modernization of the transportation facilities of the Leningrad area. Prices refer to 1985*

	Project	Investments, billions of roubles 1995 - 2005	needed for moderniz.
1.	Modernization and enlargement of the Leningrad area ports	2.5 - 4.0	3.0 - 6.0
2.	Reconstruction and modernization of the October Railway	1.5	1.5
3.	Construction of the high speed railway	5.0	10.0*
4.	Reconstruction and modernization of other railways in the area	1.0 - 1.2	2.0 - 2.5
5.	Development of the highway system of the area (including the belt way and high speed roads)	1.0 - 1.5	3.0*
6.	Development of feeder roads of the area	1.2 - 1.5	3.0*
7.	Construction of a new international airport	2.0 - 2.5	2.0 - 2.5
8.	Reconstruction and modernization of existing airports	0.2 - 0.5	0.8
9.	Construction and modernization of the Mariinsky waterway system	0.3 - 0.5	0.3 - 0.5
	Grand Total	14.7 - 18.2	25.6 - 29.8*

* = approx.

Source: Leningrad Institute of Social-Economic Problems at the Academy of Sciences of the USSR.

The fact that the state was the monopolistic owner of the majority of vehicles in the Union made their use less effective in many respects, while the road network in a market economy is very dynamic and adaptive to changes in the market. Therefore we should recognize the approach which provides new economic mechanism and new policies that will serve the true needs of the nation.

FINAL COMMENT

Summarizing the problems of the European part of the USSR, one comes to the conclusion neither that the general transport capacity, service level, quality of travel and comfort for passengers, nor the transport conditions for consignees meet global standards.

The quality of service is especially poor in the Road Transport Sector. Training in modern administration, maintenance and traffic management is urgent.

Years of hard work and huge investments are needed in order to solve the shortage in transport services. This is a concern of the general public, business management, transport industry and various others involved in trade and travel within the European part of the USSR as well as between the USSR, Western Europe and Scandinavia.

Due to inflation at the moment it is very difficult to estimate the level of investments needed for reconstruction, modernization and construction of new means of communications. Some guidelines are however the estimations for necessary investments made by the Leningrad Institute of Social and Economic Problems at the Academy of Sciences of the USSR presented above.

THE NORTH AMERICAN PERSPECTIVE

Thomas D. Larson, Administrator
Federal Highway Administration
U.S. Department of Transportation
USA

INTRODUCTION

If yo go back 150 years in American history, we were just completing the last miles of our National Pike, the first major road in the United States that was supported with Federal funds. It wasn't until 50 years later that our first Federal roads agency, the precursor of the FHWA, was established in response to numerous calls for better roads. We didn't have a Department of Transportation in those days, so the Federal roads agency was placed in the Department of Agriculture - not so surprising a location given the importance of agriculture to our economy in those days, and given the importance of good roads for getting farm products to the market. I might add that the U.S. Congress provided the grand sum of $10,000 per year for the agency's expenses.

There have been many changes since those early days as modern transportation systems have grown to accommodate significant increases in the movement of both people and goods. The mobility we enjoy today could only have been dreamt about by those who laid the early groundwork for our modern road systems. In the United States, we've built over 3.6 million kilometers (2.25 million miles) of paved roads which carry 4.3 trillion passenger kilometers (2.7 trillion passenger miles) of travel annually, and 1,022 metric ton-kilometer (700 billion ton-miles) of freight traffic. We've reached the point of completion of our Interstate Highway System, a system of roads that truly is a foundation that supports much of our Nation's commerce, as well as the day-to-day activities of our citizens.

Yet, despite the many accomplishments we've seen in building our modern transportation systems, the problems we face today are in many ways just as difficult as those of the past, perhaps more difficult, as we attempt to meet the demands of a changing population in a rapidly changing world. We are faced with great demands, limited resources, and difficult choices. Freight and passenger transportation demand continue to grow, urban congestion continues to get worse, and existing facilities continue to wear out. We have billions of dollars in unmet infrastructure needs at the same time we are faced with ever-tightening budget constraints and intense competition for available funds.

On both sides of the Atlantic there is an increasing recognition that we are moving into a new era in surface transportation. We are realizing that the solutions that helped answer yesterday's problems may no longer be the way to respond to the challenges that face us today. Whether those challenges relate to transportation, to the environment, or to energy, their solution involves consideration of a wide variety of economic, financial, technological, and institutional factors. Solutions also depend on good communications; communications between individuals, between institutions, and between nations. That is why this seminar, and the cooperation and communication we hope will flow out of it, are indispensable parts of an overall framework for problem solving.

In the United States, the Department of Transportation, under the leadership of Secretary Skinner, has taken an in-depth look at our transportation policies, to take stock of what those policies are doing well and what they are doing poorly. We sought out the comments and expertise of the users and providers of transportation, interest groups, private citizens, the academic community, and transportation professionals and public officials at all levels of government. What came out of that policy review process was a Statement of National Transportation Policy which is serving as a useful guide for shaping a national transportation system to meet the needs of the next century. I'd like to take a few moments to describe some of the basic elements of that policy statement.

THE DEMOGRAPHIC AND ECONOMIC OUTLOOK

As a first step in developing a transportation policy, we reviewed demographic and economic trends expected to influence America's future mobility needs. In terms of its implications for the economy and transportation demand, the demographic future of the United States can be viewed as having two distinct phases; the next 20 years, and the period after 2010.

- Population growth is projected to be slower than it has been in the past, but transportation will still have to serve a population increase of 32 million people over the next two decades.
- Another factor expected to increase the demand for transportation during the next two decades is the movement of our post-World War II baby-boom generation through its prime earning years. For the next 20 years, the number of people included in the middle-aged categories (ages 45-65) is expected to increase at a rate four times faster than the total population. This is likely to generate a strong demand for transportation services.
- In the period after 2010, slower population growth and the "aging of America" will become a much more significant factor in American life if current Census forecasts hold up. Between 2010 and 2030, the population growth rate is projected to slow to 0.3 percent per year, and the total population is projected to increase by 18 million.
- Slower population growth will be combined with a sharp increase in the average age of the population, with the "65 and over" age group growing eight times faster than the total population. The population share of this age group is expected to increase from 13.9 percent in 2010, to 21.8 percent in 2030.

Other trends influencing travel demand in the United States are reflected in preliminary data from our most recent National Personal Transportation Study, a national survey of trips and travel in the United States.

- Preliminary estimates indicate that nearly 60 percent of all households have two or more vehicles. The most dramatic increase in vehicle ownership is in three or more vehicle households, which rose from 4 percent of all households in 1969 to 22 percent in 1990.
- The number of households without a vehicle dropped from 14 percent in 1983 to 9 percent in 1990.
- Average annual miles driven by females grew by 50 percent

between 1983 and 1990, compared to growth in average annual miles driven by males of 19 percent.

Projections of demographic and economic growth indicate that regional and metropolitan growth patterns will play a significant role in influencing future travel trends and determining transportation investment requirements. The western and southern regions of the United States are projected to continue to be the dominant growth regions, although growth differences among regions are expected to narrow. Most growth is projected to be in metropolitan areas, which accounted for nearly 90 percent of U.S. population growth during the 1980's (between 1980 and 1990, the population of metropolitan areas increased by 20.1 million, while the U.S. population increased by 23.2 million). Metropolitan areas, with a population of just over 192 million, now account for three-quarters of our total population, with nearly 60 percent of the metropolitan area population located in suburban areas. The development of the large megalopolis, with concentrations of activity within and between suburbs, may well foreshadow metropolitan America in the 21st century.

New and far more complex patterns of trade and commerce are also emerging, and a transportation industry is developing where traditional boundaries between shippers, carriers, and public agencies are increasingly becoming blurred as a result of recent deregulation, and resulting service innovations. Our Nation's economy will increasingly be tied to the world economy and our future prosperity will depend more and more on the continued economic growth of other nations and on our ability to compete internationally. We expect global economic interdependence to become a driving force in shaping future transportation patterns.

THE NATIONAL TRANSPORTATION POLICY

With those trends as a backdrop, we developed a policy document that centered on six major themes. These themes reflect the Administration's view of the key national priorities and the appropriate Federal role. They establish a general framework for specific programs for each major mode of transportation and for each market served by transportation. The themes include:

- Maintain and expand the national transportation system;
- Foster a sound financial base for transportation;
- Keep the transportation industry strong and competitive;
- Support public safety and national security;
- Protect the environment and quality of life; and
- Advance U.S. transportation technology and expertise.

We are now giving substance to these general themes in the form of new Federal legislation and other activities for each mode – highways, urban transit, aviation, waterways, and rail. Our proposed Federal-aid highway legislation is intended to address the major policy themes by means of a substantially restructured program and special new initiatives. The proposed approach – if enacted into law – would be the first major reorientation of the U.S. Federal highway program in nearly a generation.

This landmark legislative proposal gives explicit recognition to the key role of improved mobility in support of a more productive and competitive economy by completely recasting the Federal-aid highway program for the "post-Interstate" era. The restructured and simplified program includes:

- The identification of a 241,000-kilometer (150,000-mile) National Highway System on which Federal highway investments would be concentrated in order to better serve national and interregional traffic;
- Creation of an Urban/Rural transportation financial aid program which provides unprecedented flexibility to State and local governments to cope with congestion and rural access problems;
- Strong emphasis on cost-effectiveness through management and operations;
- Innovative environmental and safety measures; and,
- Increases in the Federal commitment to the rapid deployment of new technology.

Although we are looking toward a post-Interstate construction period, we are certainly not at the end of a highway transportation economy. Our legislative proposal gives full recognition to that fact. We believe that a National Highway System is an essential part of the balanced transportation system that will be needed to serve our future social and economic needs. We believe that a

Federal perspective is needed to focus investment dollars on a select, but integrated, network of highways to ensure that our transportation system supports the Nation's future growth and productivity. While we have 50 good State highway programs, these 50 programs do not necessarily add up to a good national program. Because ours is a dynamic society, both socially and economically, we believe it is a Federal responsibility to accommodate and serve the new centers of production, population, and economic growth that have developed since the Interstate System was laid out in the 1940's. Improvements in interregional System continuity and intermodal connections to ports, airports, and to increasingly important international border crossings are also required if we are to remain strong and economically competitive.

I also cannot overemphasize the importance of increasing the level of highway investment by all levels of government and the private sector. Our legislative proposal properly increases the level of Federal funding for surface transportation improvements (a 39-percent increase for highways over the 5-year period of the legislation). It would also increase the State/local share of the costs of projects that primarily serve regional and local needs, moving the aggregate apportioned Federal share from 83 to 72 percent while enhancing flexibility and reducing Federal oversight. We also support increased flexibility to use these funds for public transport capital improvements or high-occupancy-vehicle lanes, park-ride facilities, and other measures that promote cost-effective transportation decisionmaking.

Rather than going into any more detail on our legislative proposal, I would like to say a few words about some major focus areas that, I believe, will highlight the direction American transportation will be moving as we approach the next century. These are: reducing stress on the environment, increasing the role of advanced technology, building a transportation infrastructure to meet the needs of tomorrow, and building a strong financial base for transportation.

REDUCING STRESS ON THE ENVIRONMENT

The National Transportation Policy gives a central role to environmental considerations, a role which, I believe, has received too little emphasis in the past. When the environmental movement of the 1960's swept across America, roadbuilders were happily going about their business, paying little attention to environmental considerations, and using "lowest cost per mile" as their guiding principle. To use a phrase from the Report to the United Nations Conference on the Human Environment, held in Stockholm in 1972, we had "... forgotten how to be good guests, how to walk lightly on the earth as its other creatures do".

In response to growing concern with the environment, national environmental legislation was enacted in the National Environmental Policy Act (NEPA) of 1969. As a result, highway builders began to change the way they do things. We took some difficult steps to raise our consciousness about environmental considerations. Today, we realize that transportation decisions are made in an environmental minefield, where our every step could bring serious consequences. You name it - air and water pollution, urban congestion, endangered species, wetland loss, global climate change, wasteful consumption of fossil fuels, acid rain - practically every environmental topic has a transportation connection.

I saw this clearly as Secretary of Transportation in Pennsylvania. The Blue Route, a stretch of the Interstate Highway System in Eastern Pennsylvania, is the best illustration I can find in my own experience to show how the so-called environmental restraints imposed on the highway program in the 1970's could lead to a better project. This highway is only some 32 kilometers (20 miles) long and yet for all the controversy it generated, you'd think it was a transcontinental expressway through every national park in the country. First, the State tried to build it the old-fashioned way, the way we did things before NEPA, and ended up with a short section completed in 1969 that couldn't used because, seemingly, the rest would never be built. Perhaps we ought to put an historic marker along that old section declaring it a monument to a bygone era of roadbuilding.

Thanks to NEPA - and, somewhat in retrospect, I sincerely mean thanks - the Blue Route became a testing ground in Pennsylvania for many of the environmental approaches now routine in U.S. highway development. We hired an Environmental Monitor to act as a watchdog on overall environmental issues. We formed a citizens task force, which, after countless meetings, recommended reducing the size and scope of the expressway and incorporating mass transit. The task force was probably as shocked as anyone when we took its advice. We lowered the design speed to reduce required earthwork and parkland taking and reduced the lanes from six to four, with room for high-occupancy-vehicle lanes when they are needed. We also integrated park and ride lots for transit routes and provided for possible express bus service and priority transit entrance ramps. We worked carefully with Swarthmore College, a long-time Blue Route opponent, on noise and light pollution strategies.

When I look at the Blue Route, now nearing completion, I can't help but conclude that this is a much better highway than it would have been as originally conceived. It fits better with its environment. Frankly, we wasted a lot of time and money trying to build that original version of the Blue Route. It was only when necessity - namely NEPA, a determined group of opponents, and the courts - caused us to embrace an environmentally sensitive approach that we succeeded in developing a sound, workable solution for a much-needed transportation facility.

The same kind of story can be told about many highway projects in many parts of the United States. The lesson from those stories is that, by almost any standard, the highways we ultimately built are better suited to their location than those originally planned. The environmental lessons we learned are reflected in our National Transportation Policy which recognizes the importance of environmental considerations and supports measures to protect our air, water, and land resources and the long-term health and well-being of our citizens. We believe that stronger steps need to be taken to reduce vehicle emissions as provided for in the Clean Air Act Amendments recently enacted by the Congress.

We have accomplished much in improving transportation energy efficiency and reducing transportation pollutants since the mid-1970's. Lead usage in gasoline has dropped by 99 percent since 1975. Our Environmental Protection Agency estimates that over the past 10 years transportation-related emissions have been reduced by 38 percent for carbon monoxide, 36 percent for hydrocarbons, and 15 percent for nitrogen oxides. These reductions have occurred despite significant increases in travel. Yet, despite these accomplishments, many aspects of transportation continue to produce negative side-effects on the environment, including noise and congestion, and emissions of carbon monoxide, carbon dioxide, and a variety of other pollutants that contribute to smog, threaten health, and add to the amount of greenhouse gases in our atmosphere. We clearly need to keep seeking ways of reducing transportation's negative environmental effects. The Clean Air Act Amendments will move us in that direction. The challenge we face is to clean up the air we breathe without sacrificing the goals of mobility, industrial competitiveness, and economic well-being.

Transportation and its interrelationships with the environment will surely continue to be one of the most important areas of concern for transportation builders and planners as we move toward the 21st century. We need to recognize these interrelationships, and we need to be ready to absorb the cost of environmental enhancement rather than ignore that cost and bear the consequences of a degraded environment. The Administrator of our Environmental Protection Agency, Bill Reilly, recently used the example of the countries of Eastern Europe to counter the arguments of those who say that pollution control conflicts with economic growth. He said:

> "Let them see as I have seen, rivers like the Vistula in Poland, so corrosive it is useless over 80 percent of its length even for cooling machinery; let them experience sulfur dioxide levels in Cracow, where 500-year-old statues and monuments have crumbled in just 40 years; let them see the high rates of infant mortality, lung disorders, worker absenteeism and premature deaths, the vast land areas contaminated by heavy metal pollution. Poland's Environment Minister Kaminsky estimates that environmental contamination represents a drag on Poland's gross national product of 15 percent. Policies in Communist Europe designed to stimulate economic development by foregoing pollution controls ended up by wrecking the economy and also ravaging the environment."

While we haven't reached that level of environmental degradation, many areas in the United States are faced with serious pollution problems, and many of these problems are related to the automobile. We need to recognize this, and we need to recognize that transportation, part of society's solution kit, is also part of the problem. We need to seek new solutions that bring all the costs of transportation decisions into the calculus.

The Department of Transportation, in cooperation with the Department of Energy, is looking closely at the feasibility of making further improvements in the fuel efficiency of the automobile. We are looking at ways of expanding the use of alternative fuels and we are beginning to look at ways of pricing the use of transportation facilities to reduce congestion and make better use of the facilities we have. We are also proposing substantial increases in the funding of research and development activities that will allow us to apply the latest advances in technology and the best technical expertise in attacking environmental problems.

THE ROLE OF TECHNOLOGY

The latest and the best in technology is what we must apply if we are to meet the many challenges facing transportation today. New and emerging technology offers significant opportunities to respond to capacity constraints, safety problems, energy concerns, and a variety of other transportation problems. Advances in computerization have already changed the automobile and will surely change it more dramatically in the future. The combination of advances in computers and telecommunications may lead to the day of automated vehicles operating on automated highways, with resulting benefits in travel-time savings and accompanying productivity increases, reductions in fuel use, lower operating costs, reduced environmental pollution, and improvements in highway safety.

I believe all of these things are possible, all of these things are attainable, if we have the national will to devote sufficient resources, both public and private, to transportation research and development. It is increasingly being recognized that research and development investment levels in the United States have not been sufficient to maintain a leadership role in this area. Today's level of funding for all of the national contract research, technology sharing, demonstration projects, and special national policy studies totals less than one-half of 1 percent of our annual Federal-aid highway budget. At the same time, Western European nations and Japan are undertaking significant efforts. I applaud those countries for their efforts, but I believe it is time for the United States to step up to its responsibilities also.

Accordingly, the research and technology component of a potential future highway program will provide the expanded level of funding necessary to support a long-term aggressive commitment to improving highway productivity through the development, demonstration, and deployment of available and evolving technology for both operations and construction. Under our proposed legislation, a program would be established on a multiyear basis with the following major components:

- Attention to long-range fundamental and applied research and development activities for pavement and structure materials and construction methods, safety and traffic operations research, and improved techniques and technologies for environmental management;
- A focus on intelligent vehicle/highway systems (IVHS) programs in four key areas – advanced traffic management systems, advanced driver information systems, automated vehicle control systems, and advanced commercial vehicle operations; and
- A program for motor carrier research activities, including vehicle characteristics, human factors studies, regulatory and program analysis, and other motor carrier-related research.

An important part of our research focus will be on improving the management and operation of the transportation system. Of course, we need to make the capital investments needed to preserve and upgrade existing facilities, and we must build new or expanded facilities where they are needed, but we also need to get more transportation out of the roads we already have. Research must be focused on finding ways to operate existing systems and facilities more efficiently and it must seek out new technologies and approaches to improving system effectiveness. I firmly believe that the application of new technologies can help us deal with the problems of congestion, safety, air quality, and energy conservation, while helping us serve growing travel demand.

There are a number of examples of the application of advanced traffic management strategies already being planned, installed or operated in the United States. For example, the INFORM system on Long Island, New York, was conceived as an FHWA research and development project. Today, that system is operating in a 56-kilometer (35-mile) long, 8-kilometer (5-mile) wide corridor centered on the Long Island Expressway. Included in the system are two east-west expressways, an east-west arterial, and several north-south crossing arterials and expressways, a total of 206 kilometers (128 miles) of roadways. The system features use of variabel message signs to inform motorists of unusual congestion, and it implements appropriate ramp metering and signal control measures on the affected freeways and arterials.

The Smart Corridor in California consists of 19.8 kilometers (12.3 miles) of the Santa Monica Freeway between downtown Los Angeles and the San Diego Freeway, and five parallel arterials. Traffic data and management strategies will be coordinated among the California Department of Transportation, the California Highway Patrol, the city of Los Angeles Department of Transportation and the city of Los Angeles Police Department. Improved management of the corridor will be accomplished by linking together the existing traffic control centers operated by these agencies and developing a common database of information that will promote the coordination of strategies such as ramp metering policies, parking enforcement, signal timing, and detours around congested areas of incidents.

Advanced traffic management strategies being considered as part of the Smart Corridor project include:

- Coordination of ramp meter signals with the traffic signals at the intersection of local streets;
- Use of changeable message signs on arterials and at parking garage exits to advise motorists of the best routes;
- Application of expert systems theory to the problems of incident detection and response;
- Freeway-to-freeway connector metering; and,
- A critical examination of various innovative mechanisms for communicating motorist information such as areawide highway advisory radio, computer bulletin boards, telephone dial-in systems, commercial television, and videotext.

Another example is found in the Philadelphia, Pennsylvania area, along an 82-kilometer (51-mile) section of I-95. The Pennsylvania Department of Transportation plans to transform this stretch of busily traveled highway into a model 21st century urban corridor by alleviating congestion, integrating public transport and automobile travel, and by mitigating transportation-related air and noise pollution. A traffic management plan will be implemented in the first 5 years, including such strategies as incident management, alternative route improvements, park and ride lots, ridesharing, and increased alternative mode use. During the next 5 years, dubbed the construction element, an operations center will be built and various traffic management strategies implemented, such as an incident management program, ramp metering, high occupancy vehicle facilities, pavement rehabilitation and widening, additional ramps and connectors, and various public transport-related improvements. Finally, during years 10 through 20, advanced technologies will be implemented, emphasizing intermodal state-of-the-art technologies, such as integrated ramp metering and signal control, in-vehicle motorist information systems, computerized incident management, travel demand incentives/disincentives, and advanced transit systems.

Because of the great promise of advanced technology for improving the operation of our transportation system, our legislative proposal calls for a greatly expanded science and technology program, and research and development into IVHS will be a key part of that program. We fully intend for this program to be a cooperative one, involving not only the Federal Government, but also States and cities willing to participate with us, and, certainly, the private sector as well. In fact, key aspects of this program include consideration of the appropriate forms of public/private partnership and international institutional relationships for research, development and implementation. I believe these initiatives will help us focus more attention, both by the government and by private industry, on the development and application of advanced technology to transportation problems.

In his book, Power Shift, futurist Alvin Toffler speculates on the role of information technology in changing the nature of military, political and economic power. He also comments on the connection between transportation and communications technology which may lead to a significant substitution of transportation activities by activities performed using a modernized telecommunications infrastructure.

Teleconferencing, electronic transfer of paper documents, home shopping via computer, and telecommuting, whether from home or from satellite "telework" centers, are all forms of telecommunications substitution which are rapidly evolving. A recent report by the consulting firm, Arthur D. Little, concludes that such substitutions will provide at least $23 billion in annual benefits and, on any given day, will remove up to 6 million commuter automobiles from our streets and highways.

The National Transportation Policy explicitly recognizes the potential transportation benefits of communications technology. All acress America, public/private partnerships are forming to explore and develop the potential of the "electronic highway" to ease congestion, reduce air pollution, save energy, enhance employee productivity, and increase corporate profits. I believe that many, and perhaps most, of these innovations would occur eventually in response to market forces and as a result of entrepreneurial activities. However, I also believe that the public sector has a role to play in helping to accelerate the rate at which they happen, and the Department of Transportation expects to be active in this area.

Much hard work lies ahead. We need to conduct basic research and we need to carry out operational tests. Public/private partnerships will be instrumental to the success of our efforts. The Department of Transportation is committed to a nationally coordinated IVHS program. We will serve in a leadership role, working with others, including a non-profit, public/private scientific and educational group called IVHS America, to ensure the program's success. We hope to work closely with the international community as we develop our IVHS program. Many of the challenges we face, such as human factors research, communication architecture and protocols, and system standardization, are clearly global, and the sharing of experiences and research results can pay dividends to all nations.

BUILDING TOMORROW'S TRANSPORTATION SYSTEM

While the application of new technology to make our highway investments more productive can make a major contribution to helping us meet the needs of tomorrow, the National Transportation Policy also recognizes the need to meet long-term transportation requirements by building essential new capacity. We need to maintain and preserve what we have, and we must do a better job of managing what we have, but we must also anticipate and plan for expansion in our current transportation system.

As I mentioned earlier, the focal point of our proposed highway legislation is a 241,000-kilometer (150,000-mile) National Highway System which will serve as the interstate and interregional highway system of the future. By concentrating Federal funds on that system, we will get more return for each Federal dollar spent than if we continued to spread our investment over a much larger number of kilometers where the Federal interest is less clear. The 241,000 kilometers (150,000 miles) represents roughly 4 percent of our total highway kilometers, but we expect the National Highway System to carry 40 percent of total vehicle travel and some 75 percent of all commercial vehicles.

The National Highway System will be responsive to the demands of the "new economic geography" which is reshaping the location of America's economic activity. The patterns of trade and commerce that the Interstate Highway System was designed to serve are being reshaped by new forces: new demographics, new technologies, new trade patterns, and new economic opportunities. This limited, but integrated, network will serve projected heavy truck movements, reduce wasteful circuity, and provide more direct access to centers of population and economic activity. It will play a pivotal role in serving new demands and allowing the economy to respond to emerging opportunities, whether economic change is prompted by changes in consumer tastes, by the advent of new technologies, or by changes in the international marketplace.

If we are to have a highway system that will serve these changing needs, we must also recognize the need for an increase in highway investment on the part of all levels of government, and the private sector as well. Studies of highway investment requirements suggests that a 20- to 30-percent increase is called for. Much of this must come from our traditional sources of revenue – especially motor fuel and vehicle-related taxes, but these traditional sources won't be enough. We must look to additional means of raising funds for highways, including innovative ways of raising public funds, and ways to increase the opportunities for private sector participation in highway financing.

Highway user fees – tolls and taxes on fuels or vehicles – are the predominant method of generating revenues from highways throughout the world. In most industrialized countries, revenues raised by taxes on fuels and vehicles are considered to be general purpose revenues. Transportation investment levels are set by national policy with funding coming out of general revenues, while highway user tax rates are set independently of transportation investment requirements to raise general revenues or to achieve other policy objectives.

This is not the case in the United States, where highway user revenues at the Federal level, and in most States, are substantially dedicated to transportation purposes. Highway user revenues are not considered to be general revenues, and, in many cases, are even restricted by law for use in highway-related purposes. User tax rates are intended to bear an identifiable relationship to the costs incurred by users, and highway tax levels have not traditionally been set to achieve other policy objectives or to raise revenues for nontransportation programs. A recent exception is the use of a portion of fuel tax revenues for deficit reduction. In general, however, highway investment levels are determined by what is raised from highway users.

While this user-backed system of finance still provides the bulk of our highway investment funds, and can be expected to do so for some time to come, it has been under increasing pressure to keep pace with rapidly growing needs for reinvestment and new investment. The Federal Government is committed to maintaining its strong role in funding highway development. State and local governments are also expected to raise additional revenues in order to increase their relative share of highway funding. Thirty-three States have increased highway tax rates in the past 3 years, and local governments are tapping new sources as well. Yet, despite these efforts, we need to look for additional resources, especially in rapidly growing areas with heavy investment requirements. "Conventional" roles and institutional arrangements characterized by a public monopoly and highly circumscribed roles for the private sector have served us well in the past, but they provide little incentive for the innovation in finance and production which may be essential if we are to respond to current challenges.

Toll financing has been one avenue of financial relief, with 8,000 kilometers (5,0000 miles) of existing and over 1,600 kilometers (1,000 miles) of toll facilities currently under development in the United States, including fully private projects. We expect to see expanded use of toll financing in coming years. However, toll financing is not appropriate in all situations, and many States and localities are looking increasingly to "innovative" ways of raising revenues to expand their financial base. Broadened legal and institutional arrangements have permitted local governments to organize cooperative cost-sharing or "value capture" arrangements with private sector interests. These include property owner contributions of rights-of-way and cash, utilization of benefit assessment districts, and establishment of impact fee systems.

The "innovative" financing methods are now being mixed with more conventional financing methods in new forms of partnership between the public and private sectors. Bringing together contributions, impact fees or assessments, tolls, and Federal and State taxes on fuels and vehicles, provides a reveneu base that can be tailored to fit a wide variety of circumstances – depending on the traffic levels, property development context, availability of State and local funds, and so on.

To be effective, these new combinations of revenue sources require new forms of cooperation among public and private entities in new institutional arrangements. Typically in the United States, these new public/private partnerships involve the creation of semi-independent, quasi-governmental entities separate from

State or local governments with various degrees of responsibility for project development and finance. They are supported by a broad array of innovative public/private cost-sharing techniques. Such entities are distinct from typical State or local public toll authorities, and are deliberately structured to give them access to mixed revenue sources based on close cooperative relationships with State and local governments and the private sector.

FUTURE DIRECTIONS

While finding the financial resources necessary for building tomorrow's transportation system has never been more important, perhaps the most critical task that confronts us is the task of setting realistic goals about what can be achieved. Citizen expectations about the future of transportation and the possibilities of new technologies are very high. Yet, there are no easy solutions to the problems we face. The problems of congestion, air pollution, and so on, will be with us for many years to come. We can and must confront them, and we must do what we can to see that the latest technologies are applied to their solution. But, we also must recognize, and the public must recognize, that we are constrained by the technology that is currently available, and we are constrained by limits of the public budget. And, we will be working within these constraints at the same time that travel demand is reaching unprecedented levels.

There is no single solution to these problems, but we can devote more resources to basic research and to operational tests of new technologies, we can promote more private sector participation in building and operating the transportation system, and, perhaps most important of all, we can target our investments on meeting the most essential transportation needs. I believe we also must examine new methods of managing transportation systems, even if those new methods are controversial. The consequences of one such method, peak-period or congestion pricing for use of urban highways, are being debated in the United States right now. I believe that we may find that congestion pricing, or other market-oriented approaches, may be an important part of the solution to the problem of urban congestion.

As I reflect on the title of this book, I see many opportunities and great possibilities for the future of transportation and communication. Advances in both areas are helping to draw the world closer together and are offering the promise of a better quality of life as we approach the 21st century. Many challenges must be faced if we are to realize that promise, and much work remains to be done. We must seek to meet environmental, energy, and other national goals, while at the same time making the investments that will allow transportation to meet mobility needs. Increased emphasis on research and development can help us meet these challenges, but, it will also take the concerted efforts of all involved in transportation, at all levels of government, and in the private sector. Most of all, what is required is a vision of the future that says these things are possible. A vision allowing us to keep what is best from the past, and a vision that allows us to find new and better solutions to future transportation problems. To borrow the words of the late Dag Hammarskjold: "Only he who keeps his eye fixed on the far horizon will find his right road".

THE JAPANESE PERSPECTIVE

Kazuhiro Yoshikawa
Kyoto University
Japan

INTRODUCTION

Japan has a long history of lacking of the concept of "road". The Japanese word corresponding to the English word "road" is "道(*michi*)". The word "*michi*" comprises of two syllables "*mi*" and "*chi*", each of which has its original meaning in Japanese. "*Mi*" is a prefix to "*chi*" and has no substantial meaning other than attaching a sense of courteous and significance to the noun "*chi*". "*Chi*" means "a direction" or " directions". This combination of "*mi*" and "*chi*" makes the word "*michi*", which originally meant "areas" or "regions". It is this idea behind the word that underlies the Japanese concept of "roads".

It was in the seventh century that the word "*michi*" appeared in Japanese literature. This word can be found both in *Man'yoshu,* the first Japanese anthology and in *Kojiki,* the first Japanese chronicle edited by emperors. In prehistoric times, Japanese did not have letters. After Chinese literature was transmitted to our country, Japanese did not have letters. After Chinese literature was transmitted to our country, Japanese tried to define phonetic representations of her own language by borrowing the pronunciation of Chinese characters. "*Michi*" was expressed by "美知" of Chinese characters. The character "美" designates "beauty", and "知" means "knowledge". Thus, "美知" can be interpreted as "to know beauty" or "to know reality". Of course, "美知" is nothing more than the phonetic symbol of "*michi*", since the original meaning of the characters were neglected in their trials of representing the Japanese language. However, this expression "美知" may, I believe, well express the original Japanese concept of "*michi*".

It is well known that the English word "method" came from the ancient Greek words "meta" and "hodos". "Hodos" means "roads", and "meta" is a kind of phonics which means "after", "along with", "beyond", etc. The combination of these two words make-up the word "meta-hodos", which originally meant "beyond roads" or "after roads", meaning "method" in English. One who is seeking a way to solve a certain problem may come to find by chance a plausible direction to which one should go. There is no road before one goes, but, certainly, there is a way after one goes. If many followers take the same way, this way eventually becomes a road (a method). There is no way of knowing the road before one begins to move. We only know, ex ante, a direction to which we may go. Thus, the idea underlying the word "michi" is very akin to the concept of "method(s)". As there is no unique way to accomplish one's target, there is also no unique road to reach one's destination. For Japanese, a "michi" does not simply mean a physical "road" connecting ones' origins and destinations. Rather, "michi" is the area of developed logistical systems which may sustain the movement of commodities, money, persons and information between regions, and which may catalyze the amalgamation of different cultural, religious, economic, and social experiments in local areas.

Given the geo-polical conditions, relatively isolated from the rest of the world, Japanese economy, society and culture has uniquely evolved in auto-catalytic manners. The concept of *michi* has also evolved hand-in-hand with the transformation and development of Japanese society. Nevertheless, it was in the Edo Era that nation wide road systems were established in our society. As to be described later, the roads in that era were nothing more than trails for pedestrians. Thus, Japan largely lagged behind Western nations in the development of modern road systems. It was, astonishingly, during the period of rapid economic growth after World War II that intense investment was made by the government to establish modern highway systems. Although Japanese society was delayed very much in developing modern road systems, *michi* has always played essential roles in the development of our society, economy, and culture as basic logistical systems sustaining interactions among people, commodities, information and knowledge.

For many countries, the growth or decline of regions has crucially depended on the technological state of their logistical systems, especially transportation and communications. The future states of transportation and communications of a nation depend jointly upon its own historical trajectory of development and on the future of socio-economic and technological conditions. The first part of this paper, therefore, is retrospective. My retrospection covers the development of transportation and communications in Japan during the last four centuries, which can be explained by the changing concept of what we broadly conceive by the word, "*michi*". This paper tries to shed light on the relationship between post-war economic growth and the development of transportation and communications in Japan, and to explain the basic policy formulated in the Fourth National Development Plan effective in 1987. I conclude this paper by summarizing my subjective perspective on the future of transportation and communications in Japan.

TRANSPORTATION AND COMMUNICATIONS IN THE EDO ERA

In the feudal society before the Meiji Restoration, transport systems remained underdeveloped due to institutional and physical restraints imposed by the feudal government. The Tokugawa age proved to be a period of noticable cultural and institutional development. Japan was not strongly touched by the key scientific and political conceptions which laid the basis for the European society. The "great peace (*Taihei*)" permitted the Japanese to heal the wounds of civil wars and to enjoy peacefulness over 300 years. Transportation technology as well as agricultural and manufacturing technology had remarkably progessed during the Edo era.

Following the death of Hideyoshi Toyotomi, Ieyasu Tokugawa (1542-1616) began to rule in 1603. The Tokugawa family were the most cautious and defensive of Japanese rulers. Ieyasu and his successors achieved a degree of power and nation wide authority far beyond that of previous military hegemonies. The regime kept alive a dynamic tension between feudal and bureaucratic techniques and between decentralized and centralized authorities. Ieyasu showed himself anxious to establish road systems, a functional requisite for the integration of Tokugawa political system (*Baku-han Taisei*), which was based on the parallel existence of a shogunate (*bakufu*) and two hundred and fifty *daimyo* domains (*han*). He established the main road, *Tokai-do*, connecting Edo (Tokyo) to Kyoto, along which milestones were set up every "*ri*" (ca. four kilometers) distance and 53 inn towns were founded.

The road system under the Tokugawa Shogunate was further developed by Iemitsu, third Shogun. In order to fortify Shogunate control over the provincial feudal lords, the *Bakufu* introduced the *sankin-kotai* (alternative attendance) system in 1635, whereby the lords had to maintain residences in the capital which they occupied for several months every year or every alternate year. When they returned to their domains, they were obliged to leave their wives and children behind in Edo, as hostages to the Bakufu. This system put tremendous financial pressure on the feudal lords, making them incapable of accumulating sufficient military power to plot for independence from the center. Ietsuna, fourth Shogun, established the five arteries systems connecting the main *daimyo* domains to Edo, *Tokai-do, Nakasen-do, Nikko-Kaido, Oushu-Kaido,* and *Koshu-Kaido* (See figure 1), which were directly administrated by the Transport Magistrates (*Dochu Bugyo*), one of the superindetendents of the Shogunate government. These systems led to the organization of countryside road networks. The overland highway system flourished as daimyo and their retinues traveled the roads ad as hordes of common travelers passed to and fro. Shogunal, daimyo and commoners along the roads saw to the maintenance of the road system, while bridges were sometimes neglected in order to discourage hostile movement of troops. For communications, runner (*hikyaku*) kept postal services operating for *daimyo* and commoners alike. The road network then was far from the modern highway system as it is today. Japan was slow to develop the wagon and largely relied on the packhorse for light-transport (See figure 2).

For Tokugawa, Western science and technology became nothing other than hazards. As the result, he limited trade, and prohibited the entry of all Westerners, except for Dutch merchants, embarking on the so-called *sakoku* (closure of the country) policy. This isolation lasted for 220 years until in 1859 when the Tokugawa Bakufu committed to opening three ports to Russia, Britain, France, Holland and the United States. During the period of isolation, the Bakufu initially reinforced the class system and securely established the centralized feudal structure. Under this structure domestic transport and communication networks were remarkably progressed. Transportation and communication facilities increased tremendously through both the efforts of the

Figure 1. *Five Arteries Systems in the Edo Era.*

Shogunate authorities and private companies. Castle towns developed in the provinces, and markets and relay-station communities along the road were opened due to the system of alternate attendance. In the final two decades of the Tokugawa period (1850-1868), Edo (Tokyo) was already as large as London with a population of more than one million. Shipping lines emerged with great vitality to funnel goods into the great cities of Osaka, Edo, Kyoto and Nagasaki. While tonnage was limited, small vessels in large numbers were organized to perform specific functions such as navigating the short shallow rivers into the interior or running along the coast. Bulk commodities, rice, sake, vegetables and the like were transported by boat along the coast. Express lines developed between Osaka and Edo to serve the great wholesale guilds, and these in turn formed the competitive *Higaki* and *Taru* lines. Zuiken Kawamura opened up the *Higashi-mawari* and *Nishi-mawari* lines, both of which became major routes for the shipment of rice from northern Japan to Osaka and Edo. The development of coastal marine transport facilitated the movement of commodities, money, information among the daimyo domains, and, in turn, accelerated the nation wide integration of local markets.

Figure 2. A Road for Pedistrians (Tokai-do).

The nature of education was itself a prime formative element in the Tokugawa society. Confucian-based for the most part, bookish and highly moralistic, education was nonetheless taken with great seriousness, for it was a necessary element of success for merchants as well as the *samurai* class. The Tokugawa mind was very practical and pragmatic. Knowledge on reading, writing and calculation (*yomikaki-soroban)* was regarded as fundamental requisite for daily lives. Common people were provided for in so-called "temple schools (*terakoya*)", and small private elementary schools. The result, as estimated by R.P. Dore (1965), was that by the 1860's the Japanese has achieved a literacy rate of from 40 to 50 percent among males and about 15 percent among females. All samurai were literate, and the higher level of the peasantry and merchant classes had been educated to some extent. In term of literacy Japan compared with England of the same time, a surprising fact when one considers the isolation of the Japanese from outside intellectual current. The capability of reading and writing of citizens was a fundamental requisite for the development of closely interconnected communication systems. Despite of the backwardness of technology in transportation, the Tokugawa society was equipped with knowledge-intensive communication networks. In fact, "*kawaraban* (newspapers)" served as the basic means to convey news and information. Information was quickly transmitted among citizens. Japan was able to form a unified nation immediately after the Meiji Restoration because the development of internal communication and exchange network after some 200-odd years of isolation had resulted in the near completion of the foundation work necessary for Japan to become a single country.

BUILDING A MODERN STATE - ROLE OF TRANSPORTATION AND COMMUNICATIONS

Most momentous events in history have particular central themes, and that of the Meiji Restoration (1868) is the laying of the foundation for the building of a modern state. The Japanese had interpreted the scheme of building a Western-standard modern state only in the material-physical sense, and not the spiritual sense and so, despite the rapid external and formal Westernization of science, technology, education, economics, military forces and political forms, the spiritual ideas remained unchanged (Morishima 1982).

The government devoted a lot of energy right after the Restoration to the introduction of modern Western scientific technology in the spheres of manufacturing, mining, transportation, and communications. The degree to which the post-Restoration modern development of the economy could take place depended very much on the extent to which the West's modern facilities such as steamships, railroads, electric telegraphy, etc., could be introduced. Yet all of these when judged by Tokugawa period standards, demanded unimagniable vast accumulations of capital. Therefore all of these facilities could only be introduced at first through government investment. However, as this was limited and inefficient, efforts were made for private enterprises to enter these fields.

The Meiji government's intention was to avoid, by all means, the use of foreign capital. This idea came from lessons of underdeveloped neighboring countries whose political and economic independence had been violated due to foreign capital. This had served as a profound warning to Japan since for those days foreign capital often served as the advance guard of colonization or at least semi-colonization. Hence, in 1869, when the Meiji government decided to construct the railroads, i.e., the Tokyo-Kyoto trunk line, and the Tokyo-Yokohama and Biwako Lake-Tsuruga Port branch lines, if did so without the aid of foreign capitals. The self-financed construction of railroads helped Japan avoid the evils that were experienced in other developing countries where the bulk of the railroads were built and managed by foreigners who had been granted special concessions.

In the very beginning when modern equipment was totally foreign to the Japanese people, there was nothing else to do but to rely on foreigners for building and managing the equipment and so the government hired foreign technicians. In the construction of the railroads, the main objectives of the government in hiring the foreign specialists leaders were to bring about the introduction of modern economic equipment by building and managing it, thereby presenting practical examples of the so-called "civilization and enlightenment (*bunmei-kaika*)" and to train the Japanese people in a practical manner so that they could have the scientific technology necessary for the construction and management of modern economic equipment by themselves. Under the leadership of foreign chief technician, E. Morel, the first railroad of the 25 km section between Shinagawa and Yokohama was opened on May 7, 1872. On October 14 of the same year, all the sections between Shinbashi-Yokohama (See Figure 3) were completed.

The potential ability and understanding of the Japanese was quite superior and they had great enthusiasm for mastering modern techniques. Hence, in as early as 1875, and much more by about 1877, native Japanese began to replace foreigners and more fields of activity began to be completely managed by Japanese. In the case of the Kyoto-Osaka Railway, the entire line was built by Japanese between August 1878 and June 1880. From thast time on, in all the sectors of the government railways the number of English masters decreased. All the railroad engineers were Japanese after November 1879, and after April 1883 all the masters in the railroad construction section were Japanese.

The development and spread of higher education in Japan was expected to have an important impact on the development of transportation and communication system. It soon became clear that the government was taking enthusiastic steps to fully provide higher institutions of learning and that effective practical encouragement to obtain such an education was being given by putting graduates in respectable positions. The government's first venture into university education was with a type of public institution where outstanding student from all over Japan were

Figure 3. *Shinbashi-Yokohama Railroads.*

gathered. The Southern and Eastern Colleges established under Tokugawa Shogunate were combined into the Tokyo Imperial University in 1877. With this institution a genuine university came into being whose first graduates went out into society in 1880. Table 1 shows the transitions that took place in the number of students in the different departments of the Imperial University and serve as an index to one aspect of the developing spread of higher education in Japan. We might mention in passing that the University only had four departments when it was founded, namely, law, medicine, literature, and science. Engineering was added in 1885. Besides these public institutions, private colleges were funded. The establishment of engineering departments made great contributions to the development of transportation and communications infrastructures.

Table 1. *Number of Students in Various Departments of the Imperal University*

Year End	1881	1883	1885	1887	1890	1893	1900
Graduate School	-	-	-	21	47	65	430
Law	35	55	217	218	301	416	891
Engineering	-	-	30	86	106	182	405
Literature	37	77	129	52	88	154	310
Science	85	31	43	29	97	81	65
Agriculture	-	-	-	-	485	305	288
Special Studies	-	37	73	114	-	-	-
Sub-Total	157	200	492	520	1,124	1,203	2,389
Medicine	1,355	871	726	343	188	184	491
Total	1,512	1,071	1,218	863	1,312	1,387	2,880

Note: Based on 25th Anniversary edition of Tokyo Statistical Society "Collected Statistics of Imperial University", p. 89.

TAKING-OFF OF JAPANESE CAPITALISM – ROLE OF TRANSPORTATION AND COMMUNICATIONS

The first twenty assimilations of modern scientific technology showed its main effects in increasing productivity in the areas of steamships, railroading, transportation and communications. Even among those fields, railroading had only spread to the environs of two or three large cities. Insofar as it did, it did not tie together the whole country, its contribution was small, and it did not play its full role until after the third Meiji decade.

The important reason making possible and accelerating the development of railroads were the introduction and quick development of joint-stock company system shortly after the Restoration. This was especially true in the case of the development of large-scale industries like the railroads industry. Railroad building requires vast amounts of capital and is much more suitable for the joint-stock company structure than for the individual enterprise. By stopping the tendency to bring in foreign funds and by legally prohibiting the use of foreign funds for railroad development and other social overhead uses, the government prevented the transfer of concessions to foreign business interests. This resulted in very slow initial development of railroading. However, around 1890 when these business started to bring in large profits, these revenues stayed in Japanese hands. Instead of foreigners obtaining funds from the Japanese economy which would flow out of the country, these funds were available for domestic capital formation and they became major contributions in fostering the Japanese people's ability to manage modern large-scale railway companies.

The original concept concerning the development of railroading was that they be carried out as a government-run enterprise. The lack of funds, however, limited the spread of the government's line, making progress extremely slow. The government adopted a policy of extending the railroads through private companies with the Nippon Railway Company being the first to be established in 1881. Nippon Railway capitalized 2 million yen, and built the Tokyo-Aomori line. The government guaranteed a minimum dividend in this instance, and by encouraging the nobels and capital owners to participate, the venture finally got going. It turned out that great economic success followed the opening of the Tokyo-Maebashi section in 1883. This caused a railroad fever to break out, stimulating the start of many private railways in the third Meiji decade. The development of railroading that took place after the third Meiji decade: 1) played an originating role in the rise of large-scale industries, 2) radically improved the conditions for industry with the completion of nation wide railway network that the railroad building fever of the third Meiji decade led to the remarkable development of manufacturing.

It was only around 1887 when the railroads and merchant marines so developed as modern transportation facilities began to make genuine contribution to the formation of Japan's modern state. This can easily be understood in Table 2 where the growth in railroad mileage and shipping tonnage is pictured. This growth was quite slow until about 1885 after which the situation was transformed and growth was very rapid. This growth in scale, however, does not complete the picture of the railroads' contributions to the modern development of the economy, the integration of the formerly separated sections allowed the railroads to function as one system, and, hence, show their true economic effects. Completion dates for the main lines were: 1887-Tokyo to Sendai; 1889-Tokyo to Kobe; 1891-Moji to Kumamoto; 1893-Takasaki to Naoetsu; 1894-Kobe to Hiroshima; 1901-the entire line

Table 2. Development of Railroads and Merchant Shipping

	Railroads	Steamships	
	(Miles)	Ships	Tonnage (1,000 tons)
1868	0	?	?
1870	0	35	15.5
1875	38	149	42.3
1880	98	210	41.2
1885	351	461	59.6
1890	1,698	586	93.8
1895	2,290	827	213.2
1900	3,855	1,329	543.4
1905	4,783	1,988	939.5
1910	5,354	2,545	1,234
1915	6,539	3,286	1,528
1920	8,475	5,810	3,047
1925	10,884	7,323	3,547

Notes: Railroad information is based on the publications of the Tokyo Statistical Society: Collected Statistics of Imperial Japan *(May 1928) p. 98. Shipping figures were constructed from The Bank of Japan's Statistical Section's Primary* Statistics on Japanese Economy since 1868, *p. 120.*

from Tokyo to Shimonoseki; 1905-the Ohu trunk line; 1911-the Chuo trunk line.

The growth of the merchant marine also made major contributions to the great progress in the development of Japan's modern economy through the opening of overseas routes, especially in breaking the ground in export trade routes. Before the development of the Japanese merchant marine, British and American ships controlled most of the foreign routes and quite naturally their shipping lane network and freight charge systems were devised to profit those countries.

When foreign routes were opened up by Japanese ships great contributions were made to the development of trade and the various disadvantages, inconveniences, and irrationalities of the former system were overcome and rectified. The Meiji

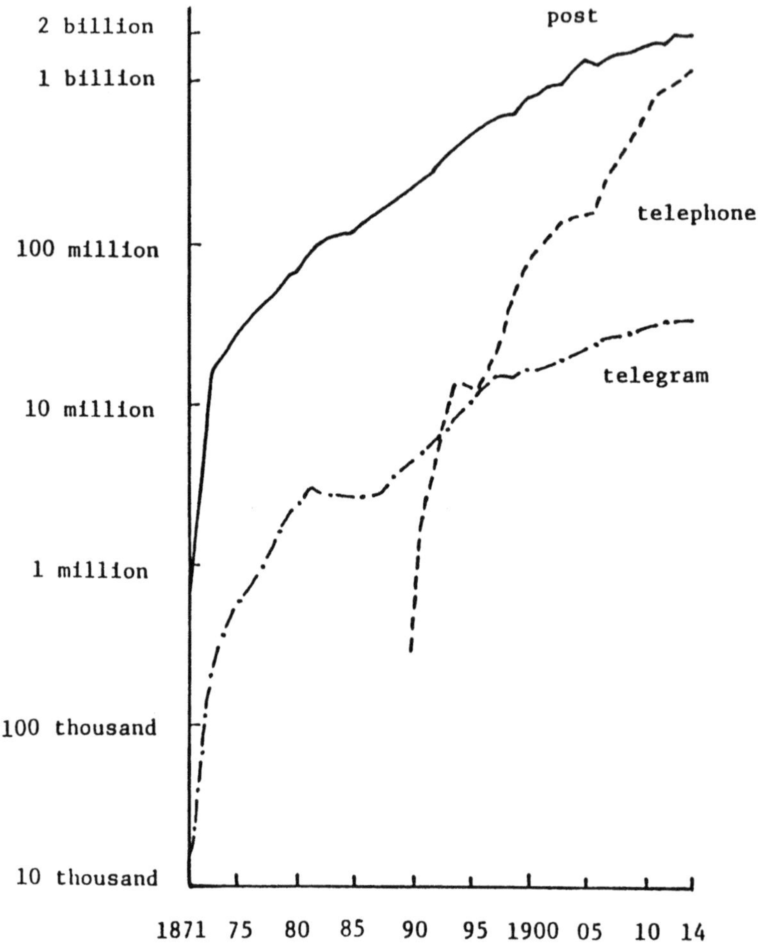

Figure 4. *The Progress of Demanded Communications during the Meiji Era. Source: "The Fifty-Years Annals of Telecommunication Development", Ministry of Post and Communications, 1921.*

government recognized the significance of the development of modern port facilities in order to promote overseas trade in the early times of the Meiji period. Introducing the Western technology of civil engineering, the significant ports in the Tokugawa Era were renovated and transformed into modern ports equipped with trade-supporting facilities. The first modern port appeared in Yokohama, followed by Ube, Osaka, Nagasaki, Nagoya, Otaru, Kobe and Tsuruga. As the amount of trade increased, Yokohama, Kobe, and other ports became new trading centers.

Regarding communications, the development of modern communication systems were initiated when the Ministry of Communications was established by the Meiji government. Telephone systems were first introduced to Japan in 1890. Figure 4 explains how demanded communications had progressed during the Meiji period. Nation wide communications systems were developed in the early 1970s, more quickly than the railroad systems, and completed in the early 1880s.

Hence, the development of transportation and communication systems in the Meiji period markedly facilitated the movements of commodities, money, and labor force among regions. However, it should be noted that the regional-uneven development of the systems caused the information and technology gap among regions. The asymmetricity of information partly resulted in the acceleration of the competition among both different economic sectors and different regions and made the Japanese economy more dynamic. It was, however, the major driving force for the emergence of hierarchical regional structure.

DEVELOPMENT OF TRANSPORTATION AND COMMUNICATIONS UNDER JAPANESE IMPERIALISM

The basic reasons for the rapid unfolding of the modern economy after the third Meiji decade were also to be found in the fruition of post-Restoration efforts which finally passed the preparatory stage and began to cause rapid developments. Around 1897, this development began to be even livelier because a number of new and powerful elements were added to the existing base, and accelerated the pace of the modern economy's development. Japan's abilities to win international security and to complete successfully among the imperialist powers were far-reaching institutional and technological reforms. These reforms also started Japan on a remarkable course of economic growth and provided the means of competition in the spheres of international trade and industrial development. Although much of what took place during the economic growth in the Meiji period was done without long-range systematic planning, it was by and large premised on the slogan of "*fukoku-kyohei* (accumulate the wealth and strengthen the military)" and "*shokusan-kogyo* (foster enterprises and develop industries)".

According to the Japanese-Chinese Peace Treaty, the Liotung Peninsula was ceded to Japan. Later, Russia and Germany began their aggression against China causing a great deal of precautionary tension to be felt by the Japanese. In the name of "post-war management" there was a rapid strengthening of protective policies taken toward modern industries with an eye to national defense. This was further intensified both during and after the Russo-Japanese War (1903-1904). After the Russian War, railroads were nationalized and was to become self-sufficient for all railway demands with efforts being exerted to develop privately run factories producing railcars and alike. The government nationalized the major railroad lines all over the country (effective in 1906-07) for military and economic purposes as well as to induce foreign capital. Various results of this included a contribution to the economic development and the development of related industries through the policy of using domestically produced locomotives, passenger coaches, freight cars, and other machinery needed by the railroads.

Emperor Meiji's death in 1912 symbolically brought to a close of the first era of Japan's struggle to become a modern nation. The new emperor, Taisho, ascended the throne under circumstances which contrasted dramatically with those which had confronted his father. The basic foundation of the imperial state had been laid. Some historians like E.H. Norman (1940) was critical of the imperial government for a policy which squeezed the farmer in the interest of an imperialistically motivated state. Later studies by Lockwood (1954) and Rosovsky (1961) showed that the government did not play dominant roles in the national economy. What is most remarkable about the imperial years of Japan was the great energy shown by the small businessman, the steady improvements of transportation and communications, and the remarkable willingness of the Japanese to put money into saving even at a low standard of living. Government patronage was mainly turned towards the encouragement of steamship lines and railroads. Postal and telegraph service developed rapidly under government management.

The narrowness of the Japanese islands, the miserliness of natural resources, and the relatively low standard of living from which Japan started its economic development, meant that from the beginning of the Meiji period there had been a constant competition for livelihood. Rapid changes in Japan's economical structure after the 1880's and in the technological foundations of the economy, created acute imbalances in rates of development. Modern industrial growth was achieved in only a small portion of the total economy, so that in the 1920's this sector was in the hands of a small group of industrial companies (*zaibatsu*) which both exploited and depressed the traditional sector of the economy. The result was what economists have called a "dual economy" in which modern industry existed side by side with traditional enterprises in such a way that the latter, serving a depressed domestic standard of living, set the standard for wage and labor practices and took on the main burden of absorbing the surplus of labor.

The year 1931 stands as a major turning point in Japan's modern history. In September of that year Japanese armed forces made an inroad to southern Manchuria, committing their government to a course of direct action on the continent and, ultimately, to the rejection of the entire structure of international relations which had come into being during the 1920's. The Manshurian Incident, of course, was less of a cause for Japan's turn toward military expansion than a symptom of deep-set domestic problem.

From the turn of the century to the beginning of the Pacific War (1941-5), i.e. the period of Japanese colonial expansion, a series of railway works were done outside as well as inside of the country. After the Seoul-Inchon railway (1897-1900), a ten-year plan for the construction of a "Trans-Formosa Railway" from Jilong to Gaoxing (a distance of 400 kilometers) was implemented. Then, in preparation for the Russo-Japanese War of 1904-5, the Pusan-Seoul and Seoul-Shinuiju lines were completed. After the war, the South Manchuria Railway Company was established, and the Shenyang-Dailian line was opened. Two outlying cities of Pusan and Dailian were finally connected by a colonial railway. During this period, military work (roads and airports) were of vital importance for the development of transportation and communications. Major pre-war projects sites were either in Japanese colonies or in areas of military significance. The owners or clients of the projects were Japan's army and navy, colonial government, or special companies founded as a deliberate act of national policy. It should be noted that the Kanmon Tunnel connecting Honshu and Kyusyu islands, was fully inaugurated in 1944, when Japan was in the midst of the war. Although it was originally planned for military significance, it turned out to become the significant means to sustain interactions of post-war ages. It was really the first trial to change the topological structure of national lands.

RISING FROM THE ASHES

Due to the evils of and the eventual loss of World War II, Japan witnessed the virtual destruction or general obsolescence of most of the infrastructure that had been built up before the war. About the only thing which Japan had to turn to after the war was the reserve of human capitals. The dominant feature of the Japanese economy since postwar reconstruction has been the very rapid and sustained growth, which was based on the existing stock of a well-educated labor force together with a large supply of managerial and technical skills. In fact, Japan had a labor force comparable to that of Western Europe in quality in 1950, but with a much lower productivity due to lack of capital, advanced technology, and fast-speed transport like automobiles and air aviation. Japan was not far behind the United States in the average years of schooling for males (8.15 versus 9.68 in 1950) and for female (7.22 versus 10.01).

It should be also noted that what sustained our country to have risen from theashes were the railroads which had been developing since the day of Meiji Emperor. Government gave a top priority to the reconstruction and rehabilitation of railroads damaged during war time. Needless-to-say, the railroads made great contributions to the distribution of foods, commodities and resources under rationed economy. The Japanese economy recovered and surpassed its pre-war level around 1953 and continued into the late 1950s with a high rate of growth primarily propelled by private business investments. However, the development of the industrial infrastructure – the setting up of a network of roads and railways, the development of ports, and the securing of land and water resources – received less-than-adequate attention during this period, with the result that shortage in such social overheads gradually came to the fore.

When the Japanese economy began to grow at the high rate, government came to recognize the absence of modern highway systems as a bottleneck for economic development. In fact, even in the late 1950s, the Japanese highway network still remained poorly developed, and was far from modern highway systems. The automobile was just minor mode of travel before the war, in the late 1950s, however, motor vehicles in Japan began to dramatically increase year by year. In fact, they increased from 1.45 per 100 persons in 1960 to 23.07 in 1973. These figures represent an extraordinary 25 percent compounded annual growth rate in the total stock of cars. Clearly, the establishment of modern highway systems had become a pressing necessity.

Figure 5. *The Opening of the Highway-Age (Meishin Expressway).*

The government set about the task of improving the road network in earnest from 1952 onward, revising the Road Law which formulates the fundamental framework for the administration of road construction, operation and maintenance. In the same year, the Special Measures Law for Construction of Arterial Motorways Construction was enacted to establish the toll road system. In 1953, Temporary the improvement of road, in order to establish the special account based on the gasoline tax which is earmarked for the promotion of road construction. A major proportion of the national budget for road development was brought under this special accounts, and its size expanded from 45.9 billion yen in 1957 to 164 billion yen two years later, a four-fold increase. The First Five-Year Road Improvement Plan was set out in 1954. In 1956, the Japan Highway Public Cooperation was founded, and the government enacted the Highway Improvement Special Means Act under which the toll expressway systems began to be constructed and managed. The institutional arrangement of special accounts and toll expressway systems underlie Japan's post-war highway improvement policies. Japan's first interurban expressway between Osaka and Nagoya, Meishin Expressway, was inaugurated in June 1965 (See figure 5) and was the most momentous event in symbolizing the opening of the highway-age. The expenditure for the construction of Meishin Expressway was partly covered by long-term debts from the World Bank, amounting to 8 million US dollars (ca. 28.8 billion yen), for a total of 119.4 billion yen. In 1966, the National Development Automobile Expressway Act was established to formulate the construction of a 7,600 kilometer network of interurban expressways with completion planned for 1983. Tomei Expressway between Tokyo and Nagoya (a total of 746.7 km in length) was completed in 1969, and, in conjunction with Meishin Expressway, served as a major skeleton of the national highway system.

RAPID ECONOMIC GROWTH AND TRANSPORTATION

The Income Doubling Plan (December 1960), visualizing the shortages in social overhead capital as a bottleneck for rapid growth, was aimed at rapid development with the result that investments in the industrial infrastructure rose from 1 percent of GNP in 1956 to 3.5 percent in 1964. Even though a major

Table 3. *Itemization of General Account Expenditures*
(Yen billions; figures in parentheses are component ratios %)

	1950		1955		1960		1965		1970		1975	
Government agencies	68.1	(10.8)	110.6	(10.9)	169.6	(9.7)	330.8	(8.9)	544.0	(6.6)	1,395.8	(6.7)
Local government finance	108.5	(17.1)	159.9	(15.7)	332.5	(19.1)	720.1	(19.3)	1,776.7	(21.6)	3,398.3	(16.3)
National defense	111.6	(17.6)	136.0	(13.4)	163.5	(9.4)	306.9	(8.1)	593.7	(7.2)	1,378.1	(6.6)
External transactions	1.0	(0.2)	11.6	(1.1)	30.8	(1.8)	19.8	(0.5)	27.0	(0.3)	18.4	(0.1)
Postwar relief	-	(-)	0.0	(0)	0.0	(0)	0.0	(0)	0.0	(0)	0.0	(0)
National lands conservation & development	93.6	(14.8)	132.2	(13.0)	294.4	(16.9)	714.3	(19.2)	1,331.1	(16.2)	2,971.8	(14.3)
Industry & economy	104.3	(16.5)	67.7	(6.6)	164.3	(9.4)	307.8	(8.3)	970.4	(11.8)	2,417.6	(11.6)
Education & culture	21.1	(3.3)	124.9	(12.3)	211.3	(12.1)	471.8	(12.7)	937.1	(11.4)	2,624.2	(12.6)
Social security	54.3	(8.6)	140.0	(13.7)	231.1	(13.3)	640.2	(17.2)	1,297.6	(15.8)	4,518.4	(21.7)
Pensions	5.1	(0.8)	88.8	(8.7)	116.3	(6.7)	157.1	(4.2)	298.4	(3.6)	755.5	(3.6)
National debt expense	58.4	(9.4)	44.2	(4.3)	26.5	(1.5)	13.0	(0.3)	287.5	(3.5)	1,102.4	(5.3)
Other	7.3	(1.2)	2.4	(0.2)	2.9	(0.2)	41.2	(1.1)	149.6	(1.8)	256.5	(1.2)
Total	633.3	(100.0)	1,018.2	(100.0)	1,734.1	(100.0)	3,723.0	(100.0)	8,213,1	(100,0)	20,837.2	(100,0)

Source: Ministry of Finance, Okura-sho Hyakunen-shi (Hundred-Year History of the Ministry of Finance), Supplementary Volume: Zaisei Tokei (Financial Statistics), 1976 edition.

proportion of this investment, amounting to 580.3 billion yen in 1964, went to the development of roads, the national railways systems also invested to the tune of 260 billion yen in ambitious schemes of its own like the building of the Tokaido Bullet-Train Line (*Shinkansen*). A significant proportion of this huge investment in national railways came out of funds for the fiscal loan and investment program.

The nature of public spending was of great significance for fiscal operations and Table 3 gives a breakdown of fluctuations in annual spending. It can be seen that major expenditures were in the areas of governmental agencies, local government finance, defense-related items, national land preservation, and development of public works, industrial economy spending, education, culture, social security, pensions, the national debt and so on. Of these, the public investment in national land preservation and development, mainly, was the highest in the first half of 1960s while, in the era of rapid growth, social capital in the form of roads, harbors, and the Bullet Train (Shinkansen) created external economies for industry. From the latter half of the 1960s onward, social security spending was expanded, reflecting the welfare emphasis of the Sato Cabinet's policies.

By the late sixties, the problem of shortage in the social overhead capital had been compounded by the so-called "strains" of rapid economic growth i the form of pollution, urban congestion and rural depopulation, and so on, forcing a diversification in policy goals. Be that as it may be how much social capital was amassed, given that spending in this area peaked during the first half of the 1960s. Although Table 4 do not show striking increases in the total extent of roads, the increase in paved roads and the development of super highways since the latter half of the 1950s have been spectacular. The influence of these two factors were tremendous in that they increased the potential for automobile transportation, accelerated the spread of the automobile, and cursed the automobile industry to develop. Moreover, their effects spread out in diverse ways as the opening up and improvement of roads made it possible to locate industry in new areas, to transport raw materials and manufactured goods, and to expand the distances over which workers could commmunicate. Table 5 summarizes the sources and uses of the funds, and shows the total as a promotion of GNP from year to year. From the very beginning, the amounts involved here gradually increased as a proportion of GNP as long as economic growth continued.

Table 4. *The Extension of Roads (figures for paved roads shown in parentheses) (km)*

	Total	Super-highways	National roads	Pre-fectural roads	Municipal roads	Private drives
1936			8,609	109,100	788,294	
			(1,197)	(2,604)	(4,593)	
1950			9,322	126,122	-	
			(1,894)	(4,554)		
1955			24,130	119,937	797,038*	
			(4,157)	(5,740)	(8,051)	
1960	961,914		24,918	122,124	814,872	
	(26,694)		(7,187)	(8,978)	(10,529)	
1965	984,934	181	27,858	128,513	836,382	415
	(61,521)		(14,244)	(16,241)	(30,855)	
1970	1,014,600	649	32,818	121,180	859,953	572
	(151,717)		(25,792)	(45,549)	(79,727)	
1975	1,067,643	1,615	38,540	125,714	901,775	622
	(338,347)		(35,136)	(83,213)	(218,384)	

* 1956 data. Private drives not included in totals.

Source: Ministry of Transportation, Doro Tokei Nenpo *(Highway Statistics Annual).*

The subject of railway transportation is one of concern and controversy in almost all developed countries. In most of OECD countries, a major concern is the decline in the quality of public transportation, especially railways services, as measured by routes covered, speed, frequency, and reliability. The announcement of the Shinkansen Plan in July 1958 inevitably evoked much controversy on the propriety of the promotion of railway-oriented policy. Majority of public opinions was rather negative or prudent in developing new railway networks. Inflammatory comments against the Shinkansen Plan, e.g. "Shinkansen – nothing more than anachronism", "stupid struggle of dying technology" alike, appeared very much in the press. Regardless of how Japanese felt about the Shinkansen Plan, the power of circumstance did not allow them to hesitate in constructing the Shinkansen lines, since it was quite clear that the passenger demands would have gone far beyond the capacity of Tokaido line in a few years. Thus, the

Table 5. *Sources and Uses of Treasury Investments and Loans*

	Sources (yen billions)			Uses (component ratios %)											
	Total	Treasury funds	Public bond issue borrowings	Housing	Environmental improvement	Public welfare facilities	Educational facilities	Small busi-ness	Agriculture, forestry, & fisheries	National lands conservation & disaster relief	Roads	Transportation & communications	Regional development	Basic industries	Export promotion	Proportion of GNP (%)
1953	337.4	298.9	38.5													2.23
1954	285.8	251.4	34.4													3.65
1955	297.8	246.2	51.6	13.9	7.7	2.1	4.6	8.2	8.9	7.6	3.7	12.3	8.6	15.1	7.1	3.36
1956	326.8	241.0	85.8	15.0	7.7	2.0	3.6	9.3	7.3	5.7	4.2	12.2	14.2	13.8	5.0	3.28
1957	396.8	351.6	45.2	13.8	9.3	1.7	3.1	16.9	6.2	4.2	2.7	11.4	9.6	20.2	0.9	5.00
1958	425.2	371.6	53.6	15.4	9.6	1.5	3.1	14.6 7.1	4.8	3.4	9.7	9.4	21.4	-	3.61	
1959	562.1	466.2	95.9	13.8	8.4	1.3	2.8	14.1	8.6	6.3	3.9	12.0	7.4	15.7	5.6	4.13
1960	625.1	506.8	118.3	12.6	9.1	1.7	3.4	12.5	7.0	6.4	4.4	14.6	7.0	13.4	7.8	3.86
1961	830.3	666.2	164.1	11.6	10.2	3.0	2.6	13.7	6.4	5.6	5.7	13.4	8.0	10.7	9.3	4.18
1962	951.3	759.1	192.2	12.8	10.3	3.0	2.5	13.8	5.6	5.0	6.5	14.2	7.7	12.7	5.9	4.39
1963	1,209.2	950.0	259.2	12.6	10.2	2.9	2.5	12.4	5.7	3.8	8.0	16.0	7.7	10.6	7.5	4.73
1964	1,403.5	1,128.5	302.0	12.6	11.3	3.1	2.8	13.2	6.1	4.0	7.8	14.4	8.1	8.5	8.1	4.73
1965	1,776.4	1,339.8	436.7	14.4	11.5	3.3	2.7	12.8	5.8	3.8	8.0	13.9	6.4	8.9	8.5	5.41
1966	2,085.4	1,471.6	613.8	15.4	11.6	3.3	3.0	13.3	6.1	3.5	9.0	12.7	5.2	7.8	9.1	5.43
1967	2,496.8	1,887.4	609.4	16.0	11.1	3.2	3.0	13.8	5.9	2.6	9.9	13.3	4.7	6.6	9.9	5.51
1968	2,783.3	2,238.1	545.2	16.3	11.5	3.2	2.3	14.4	5.7	2.2	9.7	13.2	4.4	6.6	10.5	5.22
1969	3,180.5	2,665.5	515.0	17.3	11.3	3.1	2.4	14.9	5.5	1.7	8.8	12.7	4.3	5.9	12.1	5.11
1970	3,799.0	3,301.7	497.3	19.3	11.6	2.8	2.2	15.4	5.0	1.6	8.6	13.2	4.0	5.7	10.6	5.20
1971	5,008.7	4,338.1	670.6	20.2	12.1	2.7	2.3	15.4	5.1	1.4	8.2	13.2	4.2	5.4	9.8	6.14
1972	6,037.1	5,408.6	629.2	20.4	14.0	2.7	1.9	14.5	4.8	1.9	9.4	12.2	3.9	4.7	9.6	6.37
1973	7,413.4	6,983.4	430.0	18.1	16.4	1.9	2.0	14.8	4.6	2.3	9.4	13.1	3.9	3.6	8.9	6.41
1974	9,457.8	9,058.0	399.8	19.7	16.4	3.1	2.5	15.5	4.1	1.0	8.7	13.6	3.6	3.0	8.8	6.94
1975	11,465.5	11,001.6	463.9	21.4	16.7	3.4	2.9	15.6	4.1	1.2	1.0	12.7	3.3	3.0	7.7	7.67
1976	12,403.1	11,565.5	837.6	22.7	15.9	3.7	2.4	16.6	4.8	1.1	7.6	11.4	2.8	2.8	8.2	7.34

Uses of funds for 1961 and after calculated from the original budgets.

Sources: Bank of Japan, Hono Shuyo Keizai Tokei (Hundred-year Economic Statistics of the Japanese Economy), and Ministry of Finance, Zaisei Tokei (Financial Statistics).

Figure 6. *The Bullet Trains (Shinkansen).*

construction works of Shikansen railroads with standard gages (ordinal railroads in Japan is with narrow gages) began under the program with completion in five years. It was unprecedented, in history of railroads construction, to realize the most advanced railway systems of 500 kilometer long with double track, while innovating new railway technology. Standardization and simplification of construction activities was made to shorten the duration of construction works. A variety of new technology, i.e., ATC (Automatic Train Control) and CTC (Concentrated Train Control) systems, fully continuous grand separation, anti-earthquake devices, etc., were adopted to assure the complete saftety of passengers. Shinkansen systems have been proud of no accidents up to now after being inaugurated on October 1, 1969.

Most interurban passenger travel has been by the efficient rail system of the Japan National Railways (renamed JRs after privatization in 1987). Its service-fast, frequent, comfortable, and inexpensiveness most renewed for its Bullet Train that covered the 550 kilometers between Tokyo and Osaka in 190 minutes. The line's sixteen-coach train (See figure 6) operated at more than 80 percent of their 1,400-passenger capacity and carried 270,000 passengers a day in 1971. Bullet trains leave Tokyo and Osaka every fifteen minutes during peak hours. In contrast to the situation that pertains among American railroads, JNR's interurban passenger service has been relatively profitable compared with its freight service.

JNR had begun to extend high-speed rail passenger service to a network that would link all major Japanese cities; plans called for a 7000-kilometer super-express network, by which all prefectural capitals will be accessible from Tokyo or Osaka on one-day return trips. As a part of this network plan is a fifty-four kilometer undersea tunnel to link the northern island, Hokkaido, with the main Island, Honshu. This tunnel was inaugurated in 1989. Despite the shift in transport demand from railways to roads during this period, which resulted from popularization of automobiles and consolidation of a national network of roads, large investment continued to be made in railways. Between 1965 and 1975, this figure has grown to 1 trillion yen per year. From 1965 to 1975, investment in roads also grew in parallel with the growth in GNP at a level of about 2.5 percent of GNP, although the growth slowed in the subsequent period to 1.8 percent of GNP. Moreover, two-thirds of the budget for roads was being directed toward sparsely populated areas with low transport demand and the inefficient overinvestment in national railways is revealed in the system's 2 trillion yen-a-year deficits. Despite of densely-developed railroad network, JNR had not enjoyed the scale of economy and, what is worse, accumulated financial deficits year by year, amounting to 30 trillion yen in 1986. In order to resolve organizational X-inefficiency and to call forth entrepreneurship in management, JNR was privatized and decomposed into seven JR (Japan Railways) companies in 1987.

TOWARD AFFLUENT SOCIETY

The higher-than expected rate of economic growth in Japan brought about a further concentration of population in metropolitan areas, aggravating the problem of depopulation and environmental pollution. Against this background, it is strongly requested to correct the regional disparities, to make more balanced use of national land by extending development possibilities over the entire nation, and to create affluent environment for humankind.

The Fourth Comprehensive National Development Plan was formulated in 1987. This plan, a guideline to build a national land for the 21st century, sets approximately the year 2000 as the target horizon, and aimed at the fomulation of the comprehensive framework for intermediated-termed regional development policies. In order to cope properly with the new socio-economic changes, v.z., internationalization of human networks, a trend toward an information network society, and aging of the society, the new plan fomulates the basic goals of regional development aimed at the formation of the multi-polar pattern of national land use through the promotion of the Integrated Interaction Policy based on the idea that expanded interaction bring the development of interregional and international sharing and collaboration of activities. The Plan also formulates that the Integrated Interaction Policy aimed at the expansion of interaction opportunities should be promoted by the following strategic projects which are endowed with the priorities for the implementation: i.e., (1) Build-up of vigorous and amenity-rich areas; (2) Development of Industry and preparation of living infrastructure for new affluence; (3) Preparation of transportation, information and communication systems for location settlement and interaction. This plan also formulates the basic directions of development for each regional blocks based on the viewpoints of the wider area, taking the characteristics and concerns of respective regional blocks into consideration, and presents an outline of the main measues.

Establishment of Nation-wide Expressway Systems

Highway systems as well as other transportation means, information and communication systems have been the leading strategic means which activate national economy and induce the improvement of human life quality. Toward the approaching 21st centrury, it will become more important to provide high quality highway networks for the movement of people and commodities in order to activate interregional and international interactions, to induce technological innovation, and to better human lives. In order to deepen the role-sharing and collaboration relationship among areas, and to attain development through competition among areas, the formation of nation wide comprehensive high-speed transportation systems should be established under appropriate role-sharing between different transportation modes.

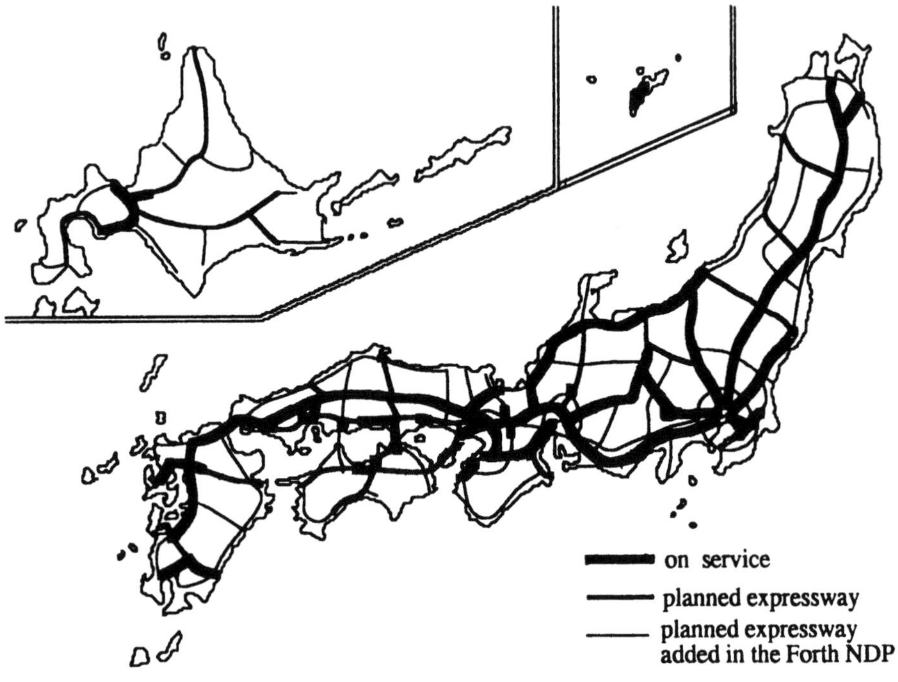

Figure 7. Plan of Trunk Road System.

It is definitively formulated in the Fourth National Development Plan to establish a total of 14,000 kilometers long of trunk road (Figure 7). By March, 1988, the section of 4,406 kilometer long of the planned network was opened. If the plan is fully implemented, people will have access to the nearest expressway within one hour, regardless of where they live in the country. "Nation wide One-day Traffic Ranges" will be established by reducing the time required for movements between major cities within three hours and the time required for access to various high-speed transportation facilities from local cities within about thirty minutes. The formation of the One-day Traffic Range is expected to vitalize regional communities through the promotion of interregional interactions.

Figure 8. *Honshu-Shikoku Connecting Bridge (Kojima-Sakaide Route).*

The three routes of the Honshu-Shikoku Connecting Bridges (Figure 8), i.e., Kobe-Naruto route, Kojima-Sakaide route, Onomichi-Imabari routes (Figure 9), are expected to play significant roles to catalyze interactions among area. The bridges connecting the main island (Honshu) and Shikoku island – they have been the dreams for habitants in both islands. Until now, twelve bridges including the great Naruto bridge of Kobe-Naruto route were inaugurated, and all sections of Kojima-Sakaide route are opened for the use of automobiles and trains. The construction work of the Tatara bridge of the Onomichi-Imabari route, possibly the longest cable stayed bridge in the world, was started in 1990 and is to be completed within this century. The Akashi Strait bridge of the Kobe-Naruto route, whose main span is 1,990 m in length, is now under construction and is expected to become the longest suspension bridge in the world.

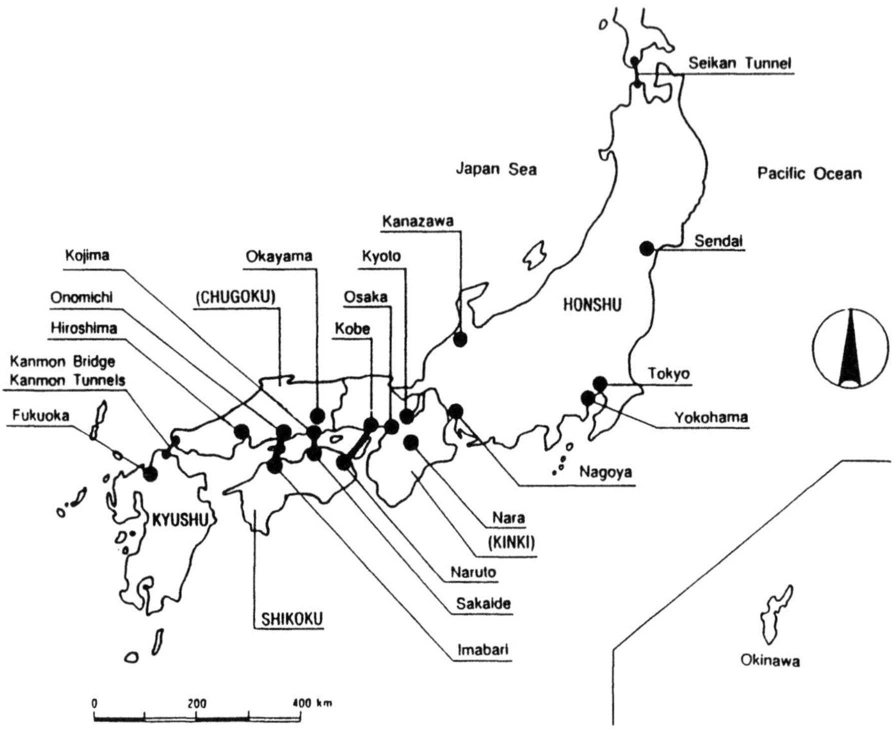

Figure 9. *Three Routes of Honshu-Shikoku Connecting Bridges.*

Establishment of High-Speed Railways

With the intentions of revitalizing local economy and of inducing balanced development of national lands, railway transportation systems of high efficiency will be provided by offering services meeting regional situations, such as stepped-up operation of trains and speed-up or improvement of connections between ordinary lines and Shinkansen lines. Of the comprehensive high-speed transportation systems, efforts will be made to prepare more efficient railway transportation systems of higher quality in line with traffic demands.

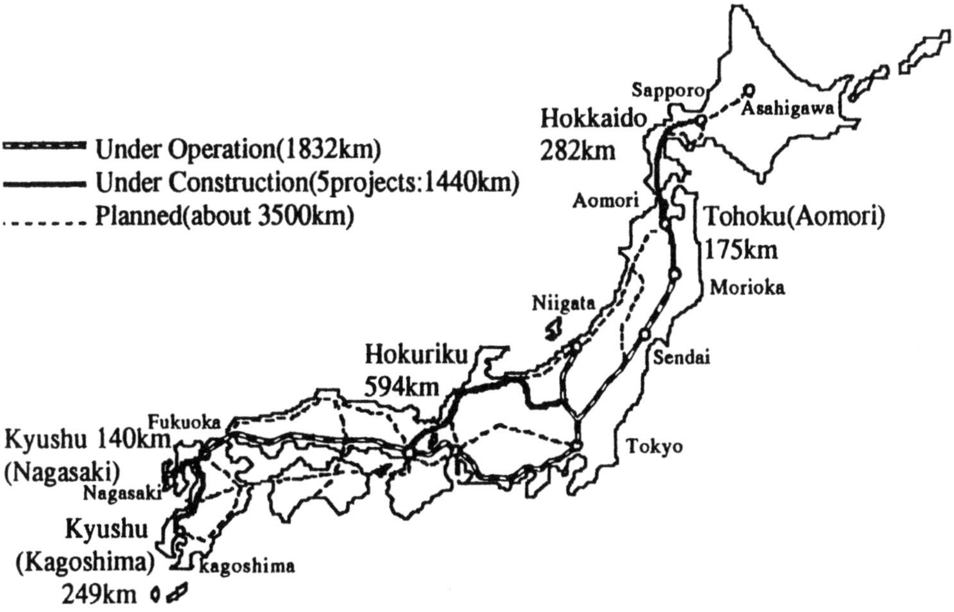

Figure 10. The Five Project Shinkansen Lines.

Up to now, Sanyou-Shinkansen, Tohoku-Shinkansen, Joetsu-Shinkansen as well as Tokaido-Shinkansen have been under operation. Accordingly, preparation and construction for the five project Shinkansen (See figure 10) lines will begin step-by-step, with taking into consideration the liquidation of the Japanese National Railways. Shinkansen will be dealt with based on the masterplan in conformity with the Law for Construction of Country-wide Shinkansen, ascertaining socio-economic trends, the results of the reorganization of the new railway operation system, technological progress, etc. Further, investigation concerning the Chuo Shinkansen line will proceed from a long-term perspective and aim at attaining a railway system of high quality, by promoting the development of new technologies such as the maglev (magnetic levitation) linear motor system (guaranteeing 500 km/hr. in speed) (See figure 11) as well as raising the level of existing technologies for the reduction of construction costs. This Shikansen would be the promising countermeasures to resolve the capacity shortage of the Tokaido Shinkansen and to reinforce interactions between Tokyo and Osaka in the future.

Figure 11. *The Maglev Linear Motor System.*

Enhancement of International Interaction Function

International movement has become routine, and improvement of convenient transportation is strongly demanded. Airports and ports which are bases for international interaction will be constructed mainly in the provincial center cities. Human exchange with foreign countries is expected to increase steadily. The number of foreign people who visit Japan in 2000 will rise to 6 to 8 million (2.3 million in 1985). Japanese travelers to overseas countries will increase to 10 to 12 million (4.9 million in 1985). Thus, the frequency as well as the quantity of international movements of passengers and freights have increased and systems capable of exhibiting the characteristics of 24 hour-operation systems will be introduced for the sake of global-scale utilization.

In order to secure smooth and constant workings of global city functions, the function of international interaction will be intensified by improvements of about fifteen airports including local airports, serving regular international flights, with trunk

Figure 12. *Plan of the Kansai International Airport and The Current Outlook.*

international airports such as those in Tokyo and Osaka as the central figures (See figure 12). For the Tokyo and Kansai metropolitan areas, flight service to rural areas will be enlarged by competing the offshore development project for the Tokyo International Airport and also by putting the new Kansai International Airport into service. The second-stage to enlarge the Tokyo International Airport has started and the necessity for the Third Tokyo International Airport will be under investigation. Further, studies promoting the overall plan of the New Kansai

International Airport will be advanced, discerning trends of air traffic demand, its profitability, etc. With an aim to enrich the integrated international air transportation services of Japan early in the 21st century, the study of countermeasures in the Chubu area, etc will be advanced, forming a clearer view of air traffic demands. The function of international interaction will be further intensified by enlarging an consolidating the principal airport of Japan such as those in New Chitose (in Hokkaido island), Nagoya, Fukuoka and Naha, in parallel with efforts to commence utilizing the New Kansai International Airport and to complete the New Tokyo International Airport.

Consolidation of Information and Communication Systems

The progress in an informatin network society overcomes time and distance among regions and enlarges the possibilities of interaction in several ways through complex networks among enterprises. Also it extends diverse information and communication media within regions. In the next decade, every person will utilize information and communications more closely and without anxiety, the technical conditions will be improved to make mutial connections possible, while promoting legal coutermeasures for protection of personal information. In order to cope with striking increases in software demand, the automation of software production processes and the development of software system programmers will be promoted. In addition to these developments, basic technologies sustaining them such as device technology as well as satellite or optical fiber communications, and the devleopment of artificial intelligence or a radio frequency spectrum must also be improved. Aiming at the consolidation of information and communication systems, the Fourth Comprehensive National Development Plan formulates the following three targets: (1) Construction of the random-access information sphere; (2) Intensification of information and communication functions in response to international integration; and (3) Formation of resilient and fully adaptive information and communication infrastructures. Together with promotional efforts to develop easy-to-use equipment, it is strongly requested to foster human competence to utilize information systems through introduction of information and communications systems in school or social education.

ESTABLISHING A MOST MODERN TYPE OF "MICHI"

Today, telecommunication systems are remarkably expanding and evolving with the aid of fiber optics, satellites and computerized switchboard systems. Airline systems are providing faster and more efficient networks for world-wide human transportation. Computers are now generally accessible for virtually all industries and many households. But, the greatest significance is, perhaps, the fact that the education system is expanding the knowledge base for citizens, thus resulting in increased efficiency in the use of computers, telecommunications and rapid transit systems.

Japan has achieved a high rate of economic growth and now boasts a world's second largest GNP, sharing 10 percent of the world's total. Japan is now endowed with an international power which exercises great influence on the global socio-economic systems, and at the same time is vulnerable to the changes in the systems that affect its very existence. Japan is shouldering its responsibilities for contributing not only to the establishment of stability of the global systems, but also to the creation of a new order for global networks. The appreciation of these responsibilities opens our eyes to the needs for a broader class of international cooperation and transactions in various fields beyond just trade-oriented relations.

The propensity toward synchronic global integration of a nation's economy and society will bring about a major world-wide redistribution of production, knowledge and information stocks, with relative decline of the Pan-Atlantic and in favor of the Pan-Pacific. In this new tide, Japan is expected to play the role of a focal point catalyzing international interactions, to form a new symbiosis of different cultures of the Orient and the Occident, and to provide comparative advantages on access to the Asian and Pan-Pacific countries with its multitude of past relations. Thus, the formation of the most modern type of "*michi*" in the Pacific Rim – a global integration of transportation, communications, resources, artistic, scientific, cultural, and knowledge networks, is a sine qua non for the future qualitative expansion of national land in the midst of new long waves, e.g., diversification of individual's sense of value, deindustrialization, global integration of industrial and knowledge complexes, and specialization and decentralization of information, etc.

The growth of knowledge base – through cultural and artistic activities, scientific researches, industrial development, administrative activities and international transactions – is a fundamental requisite for the future of the non-English speaking country, Japan. The national land should be equipped with more advanced knowledge infrastructure, providing a valuable foundation within which dominant factors for the emergence of a creative society in the 21st century may be identified. Many factors should cooperate simultaneously for the creation of a creative society. Human capitals should be expanded through international and intersectional communications. Abundant stocks of know-what, know-who as well as know-how knowledge should be intensively invested to catalyze knowledge creation. Among others, the most requested are both the human competence which facilitates the employment of the stocks of knowledge for real world purposes and the entrepreneural attitude to realize one's ambitions wherever they may come and whenever the time is right.

What made Walter E. Disney go through with his dream of creating a great leisure land for humankind was his firm belief: "If we can dream it, we can do it". For the expansion of human frontiers, there is no other time other than now that requires the ability to dream, accompanied by creative capacities and flourishing entrepreneurship.

REFERENCES

Dore, R.P., "Education in Tokugawa Japan", Berkeley, 1965.

Lockwood, W.W., "The Economic Development of Japan, Growth and Structural Change, 1868-1938", Princeton, 1954.

Morishima, M., "Why has Japan 'succeeded'", *Cambridge University Press, London*: Cambridge, 1982.

National Land Agency, "The Fourth Comprehensive National Development Plan", Government of Japan, 1987.

Norman, E.H., "Japan's Emergence as a Modern State: Political and Economic Problems of the Meiji Period", New York, 1949.

Rosovsky, H., "Capital Formation in Japan. 1968-1940", Glencoe, Ill., 1961.

PART III:

NETWORKS PERSPECTIVES

TRANSPORTATION INFRASTRUCTURE, PRODUCTIVITY AND GROWTH

Börje Johansson
Royal Institute of Technology Stockholm
Sweden

INTRODUCTION

Infrastructure consists of durable resources which can be used collectively by many firms and households. One may envisage infrastructure as a landscape of interaction possibilities for resource flows, communication, interpersonal and interfirm contacts as well as market contacts of other kinds.

Given the above formulation it is natural ask in which ways infrastructure influences the productivity of firms producing goods and services. This paper analyses the interdependence between infrastructure capital and other economic resources. The empirical study addresses questions such as (i) how is the output from industrial units determined by the combination production specific and infrastructure resources, (ii) how fast does the economic system react to changed infrastructure condition and (iii) which infrastructure variables affect productivity and growth most profoundly. The analysis is based on a production function approach, including temporal adjustments of production factors and technology. A major emphasis is placed on transportation infrastructure.

Recent Infrastructure Studies

From the 1960s and up to the end of the 1980s infrastructure investments among the majority of OECD-countries have declined as a share of the GDP in each country. The last 20 years of the same period has been characterised by a slow-down of growth as regards productivity and GDP per capita. These parallel processes have challenged many researchers during the recent decade. Aschauer (1989) and Peterson (1989) observe correlations between time series for aggregate measures of infrastructure investments and growth of economic activity. In this contribution it is demonstrated that infrastructure is a spatial phenomenon that should not be aggregated into a national capital figure in the way suggested by Aschauer. It is also shown that important information is gained when distinctions are made between different types of infrastructure.

The results in the following core based on an approach which relates to European studies by Blum (1982), Biehl (1986), and Bruinsma, Nijkamp and Rietveld (1990). In particular, the subsequent analysis has been inspired by Swedish investigations such as Wigren (1985), Andersson, Anderstig and Hårsman (1990), and Anderstig and Mattsson (1989). Two new dimensions are added here. First the model which is introduced attempts to capture the dynamics of the interdependence between economic activity and infrastructure. Second, we specify production functions which include both infrastructure capital and private (production-specific) capital.

INFRASTRUCTURE AND NETWORKS

Since the 1940s infrastructure has been referred to in analyses of economic growth. For several decades the focus was on development problems in the third world. Later the studies included lagging regions of the developed market economies. Sometimes the term "overhead capital" was used as a synonym for infrastructure. Early contributions in this tradition are Rosenstein-Rodan (1943), Nurkse (1952), Hirschman (1958) and Hansen (1965).

In this paper the interest is directed towards the built environment as infrastructure. The main focus is transportation infrastructure and pertinent accessibility characteristics. The study finds strong reasons to contemplate accessibilities to knowledge resources of different kinds. Moreover, a fundamental distinction is made between accessibility attributes within each region and each region's accessibility to other nodes.

Infrastructure in Economic Analysis

The idea of infrastructure as a growth-stimulating factor can be summarised by the following sequence of causation (compare Lakshmanan, 1985):

(i) Investments in regional and multiregional infrastructure enhance the amount of infrastructure capital.
(ii) The distribution and level of infrastructure capital influences location and capacity decisions by firms and households.
(iii) The decisions by firms and households bring about economic change processes (growth and decline) in regions. This in turn affects the over-all infrastructure investment process.

Figure 1 illustrates how a marginal cost curve is lowered from MC1 to MC2. The reduction is assumed to be caused by improvement of the infrastructure. The figure depicts cost, price and production effects. It may refer to various levels of disaggregation such as an individual firm or a production unit, a sector or the entire economy in a region.

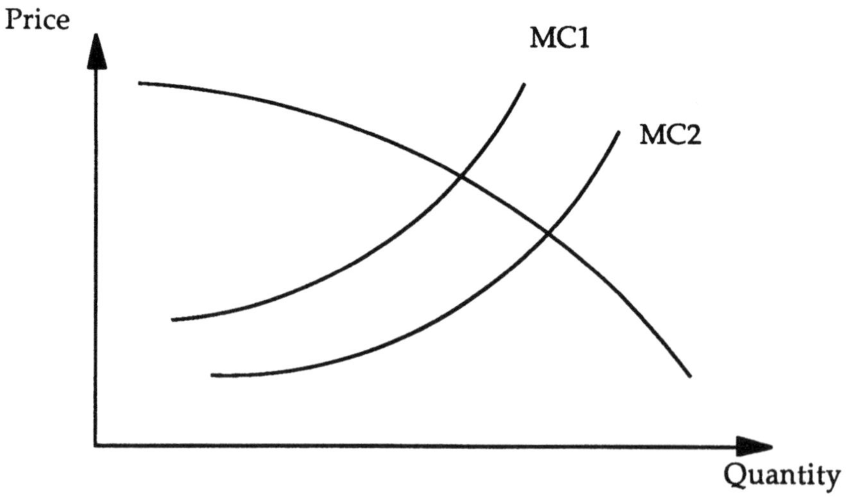

Figure 1. *An infrastructure investment which generates a shift in the cost function.*

Networks and Built Environment

Various attempts have been made to define infrastructure. From these exercises one can identify a joint core. In particular, infrastructure is durable capital, and its services have a long economic life-span, compared with the average product in the economy. Development economists have been especially generous to include intangible capital in the concept infrastructure. Examples are the health and education of the population, and social order. In Andersson (1985) the very essence is the interplay between tangible and intangible infrastructure capital.

Built environment constitutes located, durable capital. This environment can in varying degree encompass infrastructure properties. This approach relates to Youngson (1967), and it implies that the built environment and its networks are assessed with regard to systemic properties which facilitate economic interaction. This includes nearby contacts between persons and mediated interaction over long distances. In the sequel we adhere to the following set of attributes which imply that built environment is infrastructure:

(1) Infrastructure is durable capital with a fixed location.
(2) Services which can be obtained from the infrastructure has a spatial extension. Those services are normally constrained for a collective of users or consumers – or for a given service area.
(3) The utility or benefit from infrastructure services generally falls as the distance to a supply point increases.
(4) Infrastructure services are generally characterised by (i) polyvalence, and/or (ii) temporal generality, and/or (iii) systemic and network functions.

In this context polyvalence means that many different economic subjects and activities can benefit from the services. As a consequence, public goods phenomena are frequent when many households and firms simultaneously are influenced by the infrastructure.

Temporal generality signifies that a built complex can be utilised over and over again along the time axis. New infrastructure consumers can move into the built environment when old leave it.

Systems functions provide conditions for interaction and make it possible to integrate and coordinate activities in space and time.

Those networks can be intra as well as interregional.

The built environment of an urban region can to varying degrees fulfill a cohesive, systemic function. This feature is a consequence of three factors:

- Activities are concentrated in space which cause nearness and accessibility through density.
- Accessibility is built into the environment by means of intraregional transportation and communication networks.
- Abstract, temporal nearness and accessibility to other urban regions are generated by interregional networks.

Cumulative Processes

There are three main approaches to examine how infrastructure capital affects the production capability of a region. In the first approach a production function is specified, and a statistical analysis is applied to determine how the supply of infrastructure influences the productivity of other production factors. With analyses of this kind it becomes possible to assess the overall allocation efficiency and the trade off between public and private capital.

In the second approach the markets for inputs are investigated. One may then ask: how are these markets influenced by the supply of infrastructure? Reduced friction on factor markets leads to improved input combinations and an interregional relocation of labour and capital resources. Improved conditions for current inputs imply of course that production costs are reduced.

There is a natural extension of the above argument. Reduced frictions also affect output markets, and relative prices are changed. Hence, the resource allocation and activity level will be influenced in different regions. From this observation we may formulate the third approach to the study of the interdependence between infrastructure and economic performance. In this approach regional and multiregional flows of goods and services are examined (e.g. Amano and Fujita).

In all three approaches it is meaningful to distinguish between infrastructure resources within a regional node and the networks between the nodes. In this context a node signifies independent, small settlements as well as cohesive urban regions of varying complexity.

Another distinction refers to differences in adjustment speed for various variables. As a rule infrastructure capital adjusts at a slower pace than production specific capital. The fastest adjustment speed can be observed for resource flows and especially information in intra and interregional networks.

Risk capital can often be moved fast between regions as conditions change. Labour resources adjust in a slower tempo. Infrastructure adjust very slow and functions as a platform which determines the conditions for other factors. This slow pace is a characteristic which is dual to the durability, polyvalence and temporal generality of infrastructure. This is fundamental – infrastructure provides a foundation for less invariant variables.

The subsequent analysis is entirely based on the production function approach. Second, infrastructure is modelled as spatial variable. When we take away the regional specification, infrastructure loses its essence.

Consider the following production function for a sector in node r

$$Q_r^*(t)=\hat{G}_r(t)F_r^*(t)$$

$$\hat{G}_r(t)=I_r^1(t)^{\gamma_1}\dots I_r^n(t)^{\gamma_n}\,N_r^1(t)^{\omega_1}\dots N_r^m(t)^{\omega_m}$$

$$F_r^*(t)=[K_r(t+\tau)]^{\alpha}\,[L_r(t+\tau)]^{\beta}$$

$\hat{G}_r(t)$ is the infrastructure load of node r at time t. This load is determined by the nodal infrastructure variables signified by the vector $N_r(t)$, and the internodal infrastructure variables denoted by the vector $I_r(t)$. To this two vectors all other variables are assumed to adjust. Let τ^* be the time it takes for the production resources L_r (labour) and K_r (capital) to adjust to the equilibrium values $L_r^*(t)$ and $K_r^*(t)$, including the pertinent technical change. Then we have

$Q_r^*(t)=$ the equilibrium output level which applies at time $t+\tau^*$, given the infrastructure $N_r(t)$ and $I_r(t)$

$F_r^*(t)=$ the adjusted value of the production function proper which applies at time $t+\tau^*$

$K_r(t+\tau)$, $L_r(t+\tau)$ = the capital and labour resources at time $t+\tau$, where $\tau\leq\tau^*$.

One of our objectives in resent research is to measure $\tau*$. In general this adjustment period is rather long (often more than one decade). However, compared with the infrastructure, the production resource allocation adjusts on an intermediate time scale. In figure 2 we have tried to depict the cumulative aspects of the two processes simultaneously.

Specification of Variables

In this section we specify variables that play an important role in the study that will be presented in the next section. A rich set of infrastructure variables were initially collected for all Swedish municipalities. The latter are more than 280. Many of the infrastructure variables cannot statistically be connected with the spatial variation in productivity. In particular, the study shows that infrastructure capital in general must be given an accessibility description in order to explain productivity and growth differentials in a significant way.

Systems for economic and social interplay are always organised in network formations, although these may be more or less observable. The accessibility in urban environments are based on networks. Many of these extend beyond the regional boundaries.

The observations above have a general message. The infrastructure capital in a region consists of both interaction links and resources associated with different links. In this way we observe that one must always express "accessibility to a resource". The systemic nature of infrastructure arises from the fact that it includes both the resource and the pertinent network.

In the study we have two intraregional resources for which the above conclusion is important. One is "housing infrastructure" which comprises (i) every housing area and its properties, and (ii) the networks which determine the accessibility to workplaces and service and other supply points. The second example is every firm's and activity area's accessibility to different employment categories with different education and skills.

In the study the ex post selected infrastructure variables are classified in four groups. These are (i) the inhabitants' education, (ii) local and intraregional networks, (iii) regional and interregional networks, and (iv) buildings and built environment the statistically important and significant variables are presented in table 1.

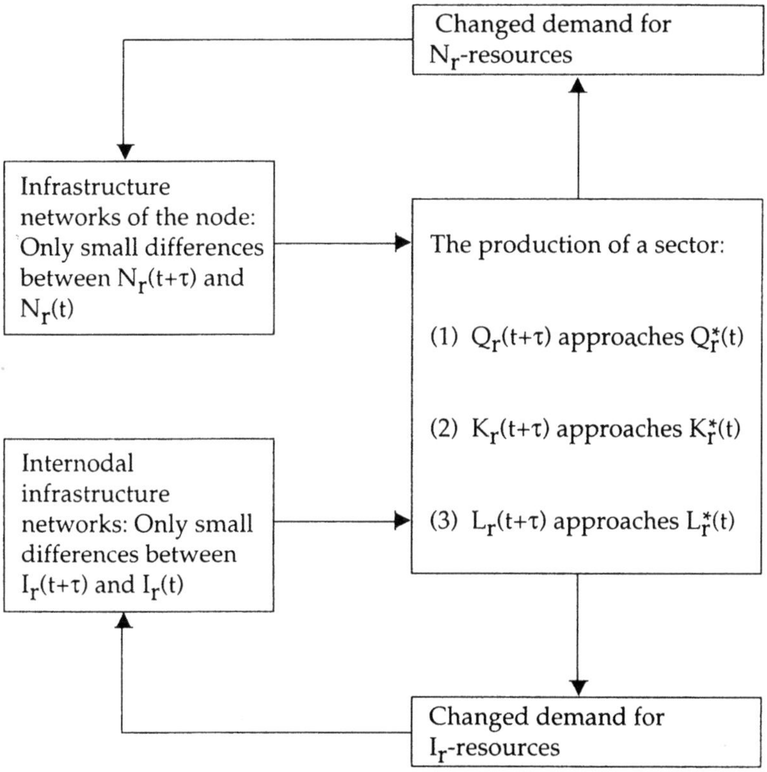

Figure 2. *Cumulative, dynamic interaction between infrastructure and production systems.*

ECONOMETRIC RESULTS

The study presented in this and the next section makes use of several models. One of these is a new version of the model type described in Andersson, Anderstig and Hårsman (1990), and Anderstig and Mattsson (1989). These model versions are static. The model presented below has a quasi-dynamic specification and is therefore called DYN, to separate it from the other models.

The formulation of the DYN-model is inspired by Mills and Carlino (1989), and Holmberg and Strömquist (1988). Important aspects of the model are presented earlier in formula. The dynamics in the formulation is related to the equilibrium output level $Q_r^*(t)$ of a sector and the adjustment process which gradually closes the gap $(Q_r^*(t) - Q_r(t+\tau))$ between the equilibrium value and the current value at each time $t+\tau \leq t+\tau^*$, where τ^* denotes the full adjustment time.

Table 1. *Infrastructure variables included in the study*

(1) EDUCATION AND OCCUPATIONS AMONG THE INHABITANTS AND THE LABOUR FORCE IN EACH MUNICIPALITY:

- College, university, graduates and postgraduates – number and accessibility
- Knowledge, information, service and goods handlings occupations – number and accessibility

(2) LOCAL AND INTRAREGIONAL NETWORKS:

- Flow capacity of the road system
- Regional accessibility in the road system
- Regional accessibility in the railroad system
- Frequency of local trains
- The capacity of public transport

(3) REGIONAL AND INTERREGIONAL NETWORKS:

- Frequency of interregional trains
- Interregional accessibility in the road system
- Airport capacity and accessibility to such capacity
- Travel time to the main air traffic hub, nearby Stockholm
- Accessibility to international terminals for freight flows
- Costal locations; accessibility to costal locations
- Accessibility to universities and research centres

(4) THE BUILT ENVIRONMENT OF A MUNICIPALITY:

- Capacity and size of the building capital
- Density and value of the municipality's buildings

The econometric model has been applied to a series of goods and service producing sectors. Here we shall discuss a set of equations estimated for the manufacturing industries. Those equations are obtained by varying the infrastructure variables included in the G-function. By presenting these equations one can show which are the core variables, in the sense that they occur in many alternative model specifications. At the same time one can assess the robustness of the estimated structures.

Infrastructure Resources for the Manufacturing Industry

Table 2 summarises the most important equations with regard to the manufacturing industry as a whole. Out of 12 comparable estimates, 6 equations have been included in table 2. In these we find the following variable frequency:

- Variable (N^1), the flow capacity of the local road system, is included in 5 equations.
- Variable (N^2), the capacity of public transport, occurs three times.
- Variable (N^3), regional accessibility in the road system, occurs three times.
- Variable (N^4), the density and quality of the built environment, is included in one equation.
- Variable (N^5), population density, is included in one equation.
- Variable (I^1), the frequency of interregional trains, occurs twice.
- Variable (I^2), the accessibility to airport capacity, occurs four times.
- Variable (I^3), road freight accessibility to international markets occurs three times.

It is noteworthy that certain parameters are extremely robust, in the sense that they maintain the same value irrespective of various alterations of the equation structure. Such an estimate is the elasticity for private (production-specific) capital. In general, all parameters in the table are astonishingly robust.

Interpretation of Local and Interregional Variables

The variables in table 2 have been organized in such an order that the first five (N^1,...,N^5) factors represent local and intraregional infrastructure networks. These are followed by three factors which are interregional and long-distance in nature.

Table 2. *Estimated versions of the DYN-model for the manufacturing industry*

VARIABLE NUMBER	EQUATION NUMBER					
	1	2	3	4	5	6
(N^1) Capacity of the road system		(0.24)	0.18	0.12	0.14	0.14
(N^2) Capacity of public transport	0.20				0.19	0.19
(N^3) Regional road accessibility	0.27				0.23	0.23
(N^4) Built environment				0.15		
(N^5) Population density	0.05					
(I^1) Interregional trains					0.02*	(0.02)
(I^2) Airport capacity	(0.004)	0.04		(0.02)	(0.02)	
(I^3) Interregional freight accessibility		0.62	0.56	0.47		
Production-specific capital	0.50	0.48	0.51	0.47	0.49	0.49
Adjustment parameter λ	0.49	0.45	0.43	0.58	0.53	0.53
F-Value	120	125	113	153	90	90

Source: Johansson et.al. (1991)

*Remark: The variables are significant above the 95 percent level, except those with * sign and those within parenthesis. The latter are significant at the 90 percent level and below this level, respectively.*

Variable (N^1) and (N^2) are especially important. They represent networks which have the mission to keep a region together, to provide a living space for households and an interaction arena for firms. For the economy the flow capacity of the road system, (N^1), is a key factor for maintaining a cohesive and robust labour market. The variable is measured in the following way:

N^1 = aV/A
a = Width times length of the main road system in a municipality, measured in square meters.
V = Driving speed of each piece of the municipality's road system, measured in meter per second.
A = The size of the built area in the municipality, measured in square meters.

The dimension of N^1 is thus meter per second, i.e., a flow characterisation. It describes the relative availability of road capital and its throughput capacity and quality. From the statistical analysis N^1 is classified as the basic factor as regards keeping a compound and robust labour market together.

Variable N^2 represents the capacity of public transport and is measured in terms of number of passenger seats supplied per day. Overall, it is uncorrelated with N^1. Still, in many municipalities it compensates for a low flow capacity of the road system.

Variable N^3 measures the accessibility to the population in nearby municipalities. It shows the same modest correlation with both N^1 and N^2. Finally, variable N^4 and N^5 describe the local environment, in terms of density and quality. This type of infrastructure components play a role over a variety of sectors, and it tends to be especially important for service and knowledge-handling sectors.

As regards the manufacturing industry, the accessibility for trucks to international contact points is astonishingly strong as a productivity predictor. With the geography of Sweden those contact points are ferry terminals and ports for container sea traffic. The dominance of this variable certainly reflects the lack of current substitutes as regards freight of goods for export and import. In particular, the higher the value per weight unit, the more important is the accessibility and quality of the road systems. This is for example revealed when the same exercise as in table 2 is repeated for the engineering industries. At the same time the essence of variable I^3 is

I^3 = Accessibility to international markets for manufactured goods.

The variables N^1 and N^2 do reflect the entire interaction quality in a region. However, they refer especially to the labour market. They are strongly correlated with the accessibility of production units to the supply of labour. This is described in table 3.

Table 3. *Simple correlations between transportation features and accessibility to labour supply*

Accessibility to labour supply with	N^2 = flow capacity of the road system	N^2 = capacity of public transport
University preparatory education	0.68	0.27
University education	0.67	0.30
Post graduate (research) education	0.68	0.31

Source: Johansson et.al. (1991).

Summarising Infrastructure Properties in Swedish Municipalities

The infrastructure variables included in our study referred to in this paper are correlated with each other in a complex pattern. Often when a municipality has plenty of certain infrastructure resources, it is also favoured by a rich fund of almost all other infrastructure variables. The overall picture may be summarised by the following four observations.

(a) Dense regions tend have values above the average as regards (i) road flow capacity, (ii) public transport, (iii) interregional trains, (iv) accessibility to knowledge-intensive labour supply.

(b) Regions with a large public sector are generally characterised by below average values for (i) accessibility in the road system, (ii) accessibility to international markets. However, in the same regions one finds above average values for public transport as well as for the frequency of local and inter-regional trains.

(c) Regions with a good supply of road capacity usually have a somewhat less rich supply of public transport and frequency of interregional trains, while the availability of local trains may be better.

(d) Regions with a high accessibility to international markets have, with few exceptions, also a rich supply of road flow capacity and a high accessibility to household with long education.

In (1) the concept of infrastructure loads $\hat{G}_r$ is introduced. One may conclude that an aggregation of the loads in each municipality to a national average $\hat{G}$ has little interest. The variation of $\hat{G}_r$-values is very high. Hence, aggregates tend to destroy all information. Infrastructure loads with regard to the manufacturing industry have 49 as its maximum value, 23 as its median. As many as 30 municipalities have a G_r-value below 15.

Suppose one wants to state in which degree there is unsufficient supply of infrastructure, from the viewpoint of the industry's productivity. Can such a question be answered by means an examination of national aggregates of infrastructure capital $\hat{G}$ and productivity, e.g. Q/L? The answer is no.

By applying an estimated equation in table 2 one can see that every municipality has its own best combination of infrastructure levels. In one municipality the payoff is largest when the built environment is improved, in another when investments are made in the road capacity. This shows that it is important not to aggregate the infrastructure resources into an aggregate and vague infrastructure capital. Our approach in which $\hat{G}_r$ has a Cobb-Douglas structure, avoids this problem. It offers both aggregate and decomposed information about infrastructure.

Next, let us investigate the rate of return to road investments, $\hat{R}_r^j$. For around 40 municipalities, R_r^j is very high. Another large set has values between 4 and 7 percent. Still another group has very low values. They fall below the one percent level. This implies that one cannot argue that there is too little road infrastructure generally, given that productivity enhancement is a guiding investment criterion. In certain groups of municipalities the payoff is high from investments in certain types of infrastructure. In more general terms – infrastructure is a spatial phenomenon, and it is based on synergies between several types of tangible networks and accessibility factors.

REFERENCES

Amano, K. and Fujita, M. (1970), A Long-Run Economic Effect Analysis of Alternative Transportation Facility Plans – Regional and National, *Journal of Regional Science*, 10:197-323.

Andersson, Å.E., (1985), *Kreativitet, Storstadens framtid* (Creativity, the Future of Metropolis), Prisma, Stockholm.

Andersson, Å.E., Anderstig, C. and Hårsman, B. (1990), Interregional Allocation Models of Infrastructure Investments, *Annals of Regional Science*, 23:287-298.

Aschauer, D.A. (1986), Is Public Expenditure Productive? *Journal of Monetary Economics*, 23:177-200.

Biehl, D. (1986), The Contribution of Infrastructure to Regional Development, *Regional Policy Division*, EC, Brussels.

Blum, U. (1982), Effects of Transportation Investments on Regional Growth, *Papers of the Regional Science Association*, 50:151-168.

Bruinsman, F., Nijkamp, P. and Rietveld, P. (1990), Infrastructure and Urban Development, Mimeo, Dept. of Economics, Free University, Amsterdam.

Diewert, W.E. (1986), *The Measurement of the Economic Benefits of Infrastructure Services*, Springer Verlag, Berlin.

Hansen, N.M. (1965), The Structure and Determinants of Local Public Investment Expenditure, *Review of Economics and Statistics*, 47:150-162.

Herzog, H.W. and Schottmann, A.M. (1989), High-Technology and Worker Mobility in the US, in Å.E. Andersson, D. Batten and C. Karlsson (eds.), *Knowledge and Industrial Organization*, Springer Verlag, Berlin.

Holmberg, I. and Strömquist, U. (1988), LOSE – Länstrafik och samhällsekonomi 1, bil. 3, *Transportforskningsberedningen*, Stockholm.

Hirschman, A.O. (1958), *The Strategy of Economic Development*, Yale University Press, New Haven.

Johansson, B. (1989), Economic Development and Networks for Spatial Interaction, *CWP-1989:28*, CERUM, Umeå University.

Johansson, B. et.al. (1991), Infrastructure, Productivity and Competition Power, in *Infrastruktur och produktivitet*, Expertutredning Nr 9 till Produktivitetsdelegationen, Allmänna förlaget, Stockholm.

Lakshmanan, T.R. (1985), Impacts of Infrastructure on Economic Development, in F. Snickars, B. Johansson and T.R. Lakshmanan (eds.), *Economic Faces of the Building Sector, Document D20:1985*, Swedish Council for Building Research,

Stockholm, 22-44.

Maclellan, D. (1990), Urban Change Through Environmental Investments, in *Urban Challenges, SOU 1990:33,* Allmänna förlaget, Stockholm, 51-76.

Merriman, D. (1990), Public Capital and Regional Output - Another Look at some Japanese and American Data, *Regional Science and Urban Economics,* 20, 4:437-458.

Mills, F.S. and Carlino, G. (1989), Dynamics of County Growth, in Å.E. Andersson et.al. (eds.), *Advances in Spatial Theory and Dynamics,* North-Holland, Amsterdam, 195-205.

Nurkse, R. (1953), *Problems of Capital Formation in Underdeveloped Countries,* Oxford University Press, Oxford.

Peterson, G.F. (1989), Lagging Infrastructure Investments; What do they Mean?, in *Urban Challenges, SOU 1990:33,* Allmänna förlaget, Stockholm, 97-114.

Quigley, J. (1990), The Quality of Housing, in *Urban Challenges, SOU 1990:33,* Allmänna förlaget, Stockholm, 39-50.

Rosenstein-Rodan, P.N. (1943), Problems of Industrialization of Eatern and South-Eastern Europe, *Economic Journal,* 53.

Wigren, R. (1985), Productivity and Infrastructure: An Emperical Study of Swedish Manufacturing Industries and their Dependence the Regional Production Milieu, in F. Snickars, B. Johansson and T.R. Lakshmanan (eds.), *Economic Faces of the Building Sector, Document D20:1985,* Swedish Council for Building Research, Stockholm, 45-54.

Youngson, A.J. (1967), *Overhead Capital,* Edinburgh University Press, Edinburgh.

Zhang, W.B. (1991), *Synergetic Economics,* Springer Verlag, Berlin.

A SYSTEMS VIEW OF INFRASTRUCTURE PLANNING

Folke Snickars
Royal Institute of Technology
Sweden

INTRODUCTION

The planning role of the Swedish municipalities is especially pronounced for the built and natural environments. In these areas there is no mixture among service production and planning roles as in several other fields of municipal activity The Planning and Building Act, and its concurrent Act for The Preservation of Natural Resources, call for comprehensive planning of the local environment, be it built or natural, as a central municipal concern in Sweden. Thus, the acts do not refer in any special form to the service production roles of the municipalities. The planning role is clearly distinguished from the other roles. The municipalities have a plan monopoly but not a planning monopoly. The implication of this is that the municipalities must decide about plans for the physical environment for there to be legally binding statements made about building permits, preservation clauses, and clearances to start building activities. The fact that there is no planning monopoly has generated an opening for the private sector to challenge the planning role of the municipalities through coordinated initiatives to plan, construct, and finance urban building projects as office complexes, shopping malls, and transportation infrastructures. However, the Swedish municipalities still have a strong position in physical planning, controlling the instruments to put forward comprehensive indicative land-use plans as well as legally binding detailed plans.

In the area of infrastructure planning there is no clear-cut role for the municipality, and a multitude of contexts in which municipal planning exists. The provision of road capital is controlled both by the state level and by the municipalities. Substantial parts of the road network are in the hands of private capital. The distinction between the local road network and the state roads is sometimes quite difficult to make. The ownership structure varies between different parts of the country. The railway network is altogether in the hands of the National Rail Administration, an agency recently separated from the state-owned railroad company. Harbours are owned by municipalities and some airports as well. The biggest airports are owned by the National Aviation Administration.

There is a difference between the position of the regional level in the different sectors of infrastructure provision. The geographical scope of the regional planning areas are different for the state road, railroad and airport planning authorities. The County Boards have traditionally had a coordinating function, a position which has been challenged by recent changes in the planning system. There county level has been concomitant with the planning areas of agencies as the National Road Administration. There is also a debate concerning the position of the Countiers and County Boards in general. Some discussants claim that the regional level in the state planning system should be totally redefined by merging the Counties into larger geographical units.

DISMANTLING THE PUBLIC PYRAMIDS

The political system of a nation is inherently domestic in nature. This is fundamentally due to the territorial paradigm in political influence and power. There have been large numbers of network nations in history. One of the most obvious ones in the Hanseatic league which generated its power from trade relationships among cities. The cities of the league were the only entry and exit point of relatively closed national territories in northern Europe. The territorially connected political power defeated the network one for many centuries after the decline of the Hanseatic league. Among the fundamental reasons for this was the enhanced possibilities of linking different parts of the hinterland of the trading cities through land transport systems. There land transport systems could be used both to transport commodities, and to exert political power over the regional settlements. The networks established through the nationalisation of the church were important in this regard in Sweden. The kings of the young nation-state of Sweden took over the clerical networks by cutting off their connections with catholic Rome, another vivid example of a network empire. It was no wonder that it took a war of thirty years to break lose from the periphery of the catholic networks and create a nation-state, stressing the common language and historical heritage of Swedes. The contermporary parallel to these network economies is the multinational corporations. They are active in many countries at the same time and accept the business rules of their host nations. At the same time the corporations have their own international networks for distributing profits.

The political changes in the European arena have fundamentally affected the way Swedish politicians think and act. Swedish politicians have gradually realised that they should take the initiative in formulating the problems to be addressed by the democratically elected bodies with respect to the internationalisation process. The private enterprises in Sweden have already chosen their strategy with respect to Europe. During the early 1990's Swedish industry was the biggest investor in Germany of all foreign countries. The Swedish inhabitants have long shown their interest in participating in the European arena by boosting international travel and thus establishing further personal contacts with continental cultures. The development implies that a larger portion than before of the Swedish electorate, especially among the younger generations, see European national borders disappear. The reopening of the eastern countries will most likely imply a changed role for Sweden, reestablishing the historical connections with the peoples and cultures around the Baltic.

The ultimate way for Sweden's politicians to get on top of this development is to change their political perspective. If this is not done, the political options will shrink considerably and adjustment to external circumstances will take over as the viable policy alternative. Such a change in perspective will also provide a way for Swedish politicians to reassess the good and bad properties of the welfare state. It is from this perspective no surprise that the social-democratic party has recently changed its long-standing views on some fundamental questions of the welfare state. The European questions might bring the youth back to politics. These young persons will hopefully be instrumental in modernising the Swedish welfare state and fulfilling the European integration.

The process of redefining the planning roles of the public sector and reorganising its activities has only started in Sweden The current tendencies are disturbing in several respects. It seems that the current political drive is towards strengthening the individually-oriented service production roles of the public sector while reducing public influence over utility services, infrastructure provision, and physical planning. Such a development is not in line with theoretical economic, and planning, analyses of the division of labour between private and public.

It is especially important for the promotion of the status of the political methods of decision-making that the regional scope is continuously redefined. The basis for the delineation of municipalities in Sweden has been the day-to-day mobility of the population. Municipalities should be formed to fulfill threshold levels of economic efficiency in the production of public services. The size of a municipality should also be influenced by the perception of the inhabitants concerning the possibilities to consult with the politicians. A municipality which has one day-time population and another night-time one, and is linked to other local areas and regions in accessing private and public services is no longer and effective geographical territory for political work. Politicians should be engaged in the solution of common problems and not collectively act as proponents of group interests. It is important that there be an independent political body at the regional level, where the region is defined so that it contains the majority of linkages. If the politicians do not take on this responsibility provate group interests will attempt to take over the public role. Several of the recent attempts from conglomerates of private enterprises in Sweden to engage in infrastructure investments can be seen from this perspective.

RETHINKING THE ROLE OF PUBLIC PLANNING

The development of any planning system is driven forward by political action. This political action is triggered by the way politicians and political parties perceive economic and social potentials or problems which warrant joint action among many persons, and organisations, to materialise or become solved. Once consensus, or majority, is reached at the political level, there is a need to create institutions which can assess the significance of the issue raised at the political level, suggest ways of amending the situation, and monitor the implementation of the policies which have been selected. The institutions which have been created to deal with one political issue will generally persist longer than the issue itself. Institutions may tend to oppose change by arguing that the issue is still not solved or that new issues can be delat with using the same institutional arrangement. There is a product cycle theory also for political products. Problems emerge, are seem as critical and demanding attention, and finally get the attention and the necessary resources to solve them. Then, similar issues are identified by other actors and solutions sought which are imspired by the way they were handled in the first case. The institutions and the policy instruments may be exported to other regions and other countries.

The political systems are still predominantly national in character. There is a trade in political ideas, and solutions, which normally can be associated with international trade relations in general. The international trade in building political and bureaucratic institutions, and formulating and implementing policies, to solve local and regional problems is still quite small. It can be argued that the public sector is not oriented to international trade in ideas and problem solutions. Rather, the standard argument is that the power over these issues should remain with the local level. Solutions should be adapted and streamlined to local conditions. The inhabitants of any municipality should vote for politicians living in the area, and they should be instrumental in solving the local and regional problems. There are reasons to believe that a development in this direction will not be conducive to a strengthening of the status of the politicians.

There is no rational reason to totally shift the decisions about joint questions of service production and planning to the private sector. In the knowledge, the need for collaboration and interaction in society will increase substantially. The idea that decisions about common matters should be made more efficiently if market mechanisms were used may be valid if reference is made to the planned economies, but not necessarily for a mixed economy of the Swedish type. Such ideas are also sometimes based on a poor understanding of the contents of the market mechanism. Non-market institutions are fundamental to the rationality of the game of transforming political ideas to practical policies.

The rules of the game are also important, more important than the way that the game is actually played by the different actors involved. In this context, the price system is a special case of procedures of negotiations between demanders and suppliers of private and public goods and services. The importance of the price system lies in the trust in its operation. If there is no fundamental trust, as was the case for a number of years in the countries in eastern Europe, a shadow price system will appear which shows the real magnitudes of demand and supply in the technical and organisational context that the political system has generated. Much in the same way that the trust in the price system comes from actions taken within the institutions which have been created to handle it, so does the political system gain its status and legitimacy from the performance of the institutions which have been established to support and develop that system. The political system is not excluded from the principle that fundamental system reform will not emerge from within the system itself.

Policians have often seen the legal system as the institutional infrastructure for political decision-making. There has also been a reliance in bureaucratic organisations shaped to implement and monitor public policies. In this context, the negotiations element is central. Price systems, legal frameworks, and bureaucratic systems are all methods to save on direct person-to-person negotiations. Such negotiations are costly in terms of economic resources and inputs of time from the individuals. It is only in very complicated decision situations that direct negotiations among actors will take place. The aim of such negotiations will always be first to create joint values and gains, and then to distribute these gains among the actors involved. The fundamental role of the public sector in this regard must be to represent those interests which cannot be articulated by the private actors. These interests are the ones of the economically and socially weak, and the interestes accruing from the position of future generations with regard to welfare and environmental sustainability.

An important element of a strategy for renewal of the public sector in its planning activity is to distinguish more clearly than before between the demand and supply role. At least in Sweden, there has been a tendency to mingle the two roles with one another both within the public sector and in the rules guiding the interrelation between the private and the public perspective. It is fundamental that the demand role stays with the public sector, and that this role at the municipal level is not merged altogether with the political actors. The politicians should create institutions and give them goals and resources, define the rules of the game, and the principles for the generation and distribution of economic resources. The demand role should be separated from the supply one in these instituions. There is no reason to maintain public monopolies on the supply side. Instead, competition should be encouraged between private enterprises concerning the production of individual or utility services.

There is nothing that excludes the possibility that there will be benefits generated from further competition in the preparation of plans for the physical environment of a city or region. The introduction of competing alternatives will ultimately increase the quality of the plan alternatives chosen. Even though the municipalities have a plan monopoly, there is no reason to exclude planning as an activity from market competition. In fact, it must be regarded as highly detrimental to planning as a professional field, at least in Sweden, that plan alternatives have normally been produced within the public sector. The fact that this has caused inefficiencies on the supply side has entailed an unwarranted questioning of the supply role of the public sector. The public sector must take the lead in challenging the private sector to increase its creativity in thinking about public sector issues. This should be seem as a strategy to reestablish planning as a self-evident element of public decision-making in Sweden as in other knowledge-oriented societies.

The legal framework of urban planning in Sweden is not conductive to shaping an atmosphere for the creation of joint values between actors. Instead, the acts controlling planning and building as well as the exploitation of natural resources mix the demand and supply roles of the public sector. Plans are to be prepared by the public agencies, and them displayed to the public as a part of the approval process. There is a system for bringing in the inhabitants through compulsory consultations with the general public before plan proposals are made. This system of participatory planning generally will bring the inhabitants in at a late stage in the process and stir up little interest. In reality, the plans will often be superseded by actors who wish to perform specific building projects. Their proposals give them an initial initiative of defining the stage for negotiations which has proved quite difficult to challenge.

Recent experience has shown that the public has had difficulties to meet these initiatives with strong arguments collected through the current planning schemes. Instead, the public sector agencies have volunteered to negotiate directly with the proposing actors about the conditions for collaboration. Contingent contracts are drawn up which presuppose that planning schemes will be approved after the project agreements have been struck. The result is often that proposals are blocked through environmental groups or other actors favouring preservation of existing situations. The legal system ties the public actor to inflexible behaviour in these contexts. The focus on process and procedure in the legal systems controlling planning weakens the role of the public perspective. In the rethinking of the role of public planning in Sweden, there is reason to look to the experience of other countries also in this regard. Otherwise, the Swedish way of performing physical planning may prove less adequate than warranted in an international context. There is reason for the Swedish planning actors to place their new planning Acts in perspective although they have only existed for a few years.

THE PLANNING OF INFRASTRUCTURE SYSTEMS

Mobility is a central concept in the understanding the current mechanisms of change. We have witnessed a tremendous increase in the mobility of production factors over the recent decades. People and goods are transported over longer distances without a corresponding increase in the time spent on travel. Households are spending almost the same amount of money on transport and communication as on housing which is a substantial change over the last twenty to thirty years. Fírms are boosting their budgets for business travel. At the same time, the increased value content of goods has entailed a slower increase in the costs of transportation for freight. The mobility of the labour force has expanded the scope of the Swedish local labour markets. There are now quite few isolated local labour markets at least in south and central Sweden. People have becöme used to spending several weeks per year away from their home regions.

This development has brought transportation up as a contemporary political issue. Its importance has increased further through the environmental problems associated with the use of petroleum-based fuels. There are two opposite ways of looking at transportation demand in this regard. One is to see the development as givning rise to a shortage in the supply of transport infrastructure. The interest then will become focused on methods of increasing the investments in transportation networks. The other way is to see the development as giving rise to an excess demand for transport. The interest then becomes focused on the pricing of transport services.

The mobility increase has been one of the factors weakening the role of the local politicians. When voters spend substantial amounts of time in other municipalities, as many members of the workforce nowadays do, it is not self-evident where the political emphasis should be placed. It is not evident, either, which actors should suggest improvements in the transportation systems connecting the formerly separated local labour markets. The result is a more complex role for infrastructure planning.

The recent development in Sweden is that the state agencies for infrastructure planning have reorganised themselves. The National Road Administration has been one example in kind. There is now a split between the planning role and the building and management role of this planning agency. The county level has been removed as the regional level both in the planning system and in the management and road construction activity. There are now five planning regions and seven construction regions. The regional borders do not coincide and the regional centres of the two branches of the National Road Administration are not identical. The current work is aimed at setting the stage for the planning activities of the new regional planning organisations and at clarifying the division of labour between the regional level and the national one. The work is interesting to follow. It is also interesting, and indeed illustrative of the case we have wished to make, how this reorganisation is connected to changes in the political situation.

The new regional planning structure of the national agencies does not fit with the political boundaries. The politically elected County Councils operate within the county framework. The politicians are not specialists in planning problems but rather in health care questions. The chance is that their competence, and legitimacy in dealing with the new regional issues, will be questioned. It is not surprising that there have been a large number of unconventional alliances between local and regional politicians, and business land trade-union leaders, in the promotion of infrastructure investments linked to broader regional milieus than the municipal arena. The tendency in Sweden is that these developments are not guided by the elected politicians to the extent that one would wish for democratic reasons. It is typical of the current situation that new coalitions are formed to block out initiatives taken by other coalitions, creating stalemate conditions where economic and social values can not be created jointly before the distribution of net benefit is claimed.

One reason for the current imbalance in the regional political system is that the change to a systems perspective has not been taken soon enough at the political level. This has shrunk the options available for planning to an unwanted level. One background of the viability of the organisational changes in the National Road Administration was the shift from an object-oriented view of the transportation network to a system-oriented view. In the latter mode of thinking the starting point is a broad perspective on transport systems and traffic formation. In some cases, the systems context may be so well-delineated that an object-oriented view can be undertaken. However, this is rather the exception than the norm. In the earlier mode of operation of the National Road Administration, the link to management and road construction further strengthened the interest in local issues. The new context is quite different. Now, the important discussion concerns the interaction among modes of transportation, and the complementarity and substitution among infrastructure networks.

It will be a challenge to see whether the political decision-making system will produce such systems innovative thinking by itself. If this is done, we might ultimately be prepared to speak about a systems view of infrastructure planning.

SOME ASPECTS ON INTERDEPENDENCE OF COUNTRIES AND REGIONS ON IMPROVED TRANSPORT LINKS

Willem Westerhuis
Director General, the International Road Federation, Geneva
Switzerland

The days when decisions regarding infrastructure were a purely national affair are over, at least in theory. The interdependence of countries and regions on improved transport links for economic growth is now generally recognized although views on which links and what modes to promote differ widely, not to mention the burning question of how to pay for it all. These issues give rise to heated debates at practically all levels of government in many countries. You can imagine the scene when these problems are debated at an international level. A good example of the complexity of major decisions involving two or more countries is illustrated by the fixed link between Sweden and Denmark. The only way to cope with expected growth in transport needs is to tackle the development of international networks on a supranational level. Why this is so can best be demonstrated by some basic statistics. The development of transport in Western Europe has taken place parallel to the evolution of the GDP. During the period 1970 - 1988 GDP on average increased by 2.5 percent per annum, freight transport also by 2.5 percent and passenger transport even by 3.1 percent. The E.C. estimates that with the introduction of the single market an annual GDP growth rate of 4.5 percent in the Community is realistic.

A very conservative estimate, i.e. based on a continued economic growth of only 2.5 percent per annum, will yet give an increase in transport of 34 percent and in international transport an even higher percentage by the year 2000. Similarly, with this economic growth, mobility is expected to be 25-50 percent higher by the year 2000 compared to the volume of traffic in 1988.

The road is by far the predominant mode of transport, this is not due to change, but reflects the natural market mechanisms and the demand for economic growth (higher standards of living). The road is simply the easiest accessible and most reliable means to get mobile. 77 percent (2.5 billion tons) of goods exchanged between E.C. regions in 1987 were transported by road.

Investments in surface infrastructure in general and in roads especially have sadly failed to keep up with traffic increases over the past 15-20 years. The results are there for everyone to see, underdeveloped urban transport systems, inefficient railway systems, serious congestion on many major road links and far more pollution than need have been. If this is not enough there is also a serious incompatibility in the various transport networks from one country to the next.

An overall view is sadly lacking. For that reason the IRF developed a plan called **AIMSE** which stands for **Advanced Integration Motorway System in Europe**. The resulting report proposes the integration and extension of existing national motorway networks in West and Central Europe in order to cope with the inevitable heavy increase in road traffic in coming decades. The project includes the improvement of existing motorways, the construction of 20,000 km of new links, urban ring roads, tunnels and bridges, the protection of the environment, high tech information systems, safety and communication equipment. It will have to be financed through a combination of public and private funds, with a large recourse to user charges wherever feasible and justified. Point of departure was to approach the transportation problem across Europe on the basis of a global evaluation of realistic needs in the long term. A guiding principle was that an unbridled expansion of road networks should be avoided through capacity improvements on present and future motorways as well as an improved integration with other available modes of transport. The fact remains however that roads will remain the predominant mode for decades to come. There is simply no alternative in sight. For example a doubling of railway capacity will only absorb one year's expected growth in transport needs.

Despite the fact that the AIMSE report gives special emphasis to the member countries of the European Community it would be unrealistic to limit its impact to the E.C. only. Especially for countries like Sweden on the periphery of Europe, transport and communication links are a vital importance. Transport time and cost need to be kept at a minimum to preserve competitiveness in the enormous market place of the new Europe. The fixed link with Denmark is not built a minute too soon, although the link by itself will not be enough to assure future prosperity. Production centers throughout the country must have easy access to the most efficient (time and cost effective) transport means possible. In many if not most cases the road or rather the motorway will be indispensable.

Throughout Europe heavy investments will have to be made in order to eliminate the effects of procrastination in the past and to prepare for future needs. Available money is far too limited to meet these needs. However the ways of financing the construction, management and maintenance of road infrastructure are manifold. From the users point of view possibilities are limited to two main procedures: tollways or freeways. The basic concept of the toll system is that the users themselves directly support expenditures. In a typical toll motorway system an average of 65 percent of the total income is used for refunding loans, 15 percent applied to maintenance, and the rest for toll collection etc. Profitability of motorways is generally assured if traffic volume is 15,000 or more vehicles per day.

To sum up IRF's position on the future of transportation in Europe as a whole I should like to stress that:

- The demand for transport and mobility will increase significantly between now and the year 2000.
- Transport infrastructure, already reaching a breaking-point in many parts of Europe is grossly inadequate.
- Optimum development and integration of all transport modes need to be vigorously pursued to prevent the economy grinding to a halt.
- Roads will continue to provide the backbone of Europe's infrastructure.
- To perfect the European road system a core motorway network, along the lines of IRF's AIMSE project must be initiated very soon.

With regard to this last point I can with due modesty say that the Autorities in Brussels have woken up to the fact that their notable preoccupation with the promotion of high-speed rail links and combined transport systems is no more than scratching the surface of the problem. Despite frequent political preference for other modes of transport, reality dictates otherwise, and in consequence Brussels is now using our AIMSE project as an incentive for a similar study to be completed by the end of this year. Unfortunately the road from theory to practice is long and winding and I am afraid it will be some time yet before concrete measures are taken. Let us hope we can still afford to wait that long. The IRF at any rate will use all its means and influence to speed up the process.

TRANSNATIONAL PLANNING EDUCATION AND THE INTERNATIONALIZATION

Klaus R. Kunzmann
University of Dortmund
Germany

TRANSPORTATION, COMMUNICATION AND PLANNING EDUCATION

Over the past decades planning education in Europe has been established as an independant discipline, independant from

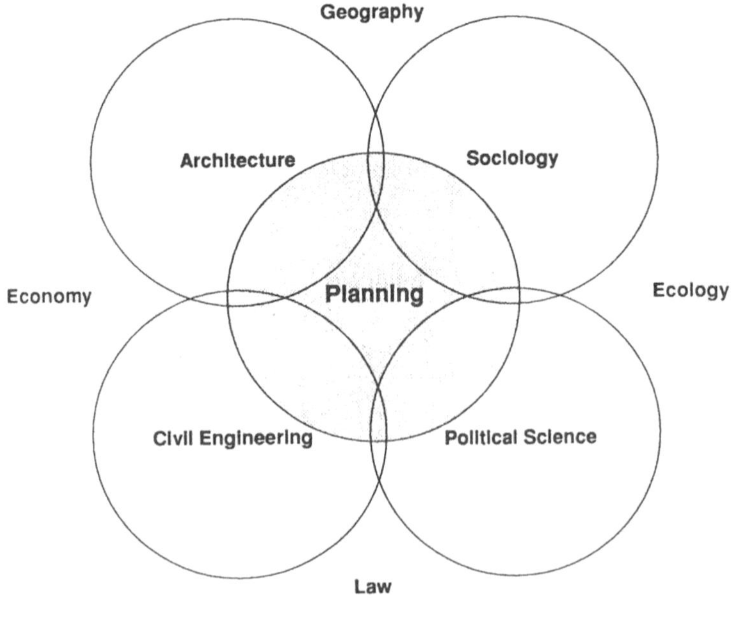

Figure 1. Planning as a scientific and professional discipline.

Figure 2. Planning schools in Europe (1990).

traditional disciplines such as geography, social sciences, economy or architecture. Piloted by the explorative efforts of English (and North American) planners and drawing on the knowledge of the traditional disciplines planning education has become both an academic and professional field of studies (see Figure 1).

In 1991, more than 80 universities in Europe offered a course in planning (Figure 2).

Although transport and communication are key variables in explaining and guiding spatial development, with only few exceptions, these variables (and this is more than the traditional system of land-use/transport interaction) seem to receive comparably low support in these planning schools.

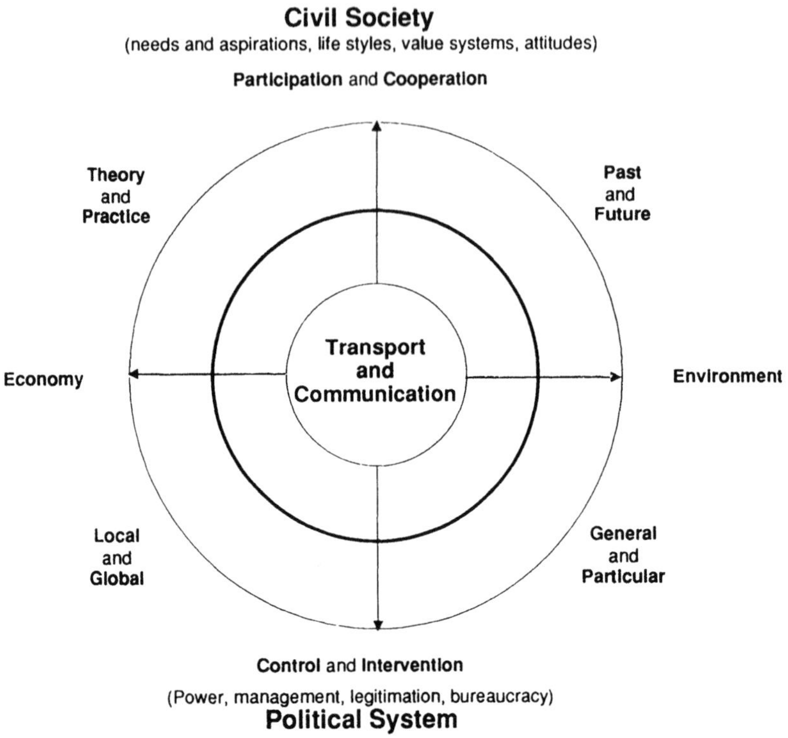

Figure 3. *The research context of transportation and communication.*

It are rather the transport courses in civil engineering departments which cover, or claim to cover, the subject area. Understandingly this is done with a strong bias on technical and quantitative aspects and aims at educating civil engineers as transportation planners. All other aspects of the complex transport and communication field are increasingly subsumed in schools of mechancial engineering (there called "logistics") and in schools of economics (transport economy), or in sociology, psychology and in communication or political sciences. In such areas the social and behavioural consequences of increasing mobility, of new communication means and uses are researched and taught.

Regrettably most planning schools seem to neglect this subject matter, whether they are focusing more on the physical or on the policy aspects of urban and regional development and planning. This happens despite the growing importance transport and communication plays in the information and communication society of the forthcoming 21st century (see Figure 3).

WHAT ARE THE TRANSPORT AND COMMUNICATION RELATED ISSUES IN PLANNING EDUCATION AND RESEARCH IN THE COMING DECADE?

Given the growing negative impacts, mobility and traffic and related energy consumption have on our world environment, the foremost aim of future urban and regional policies must be to explore ways and means of how to reduce again traffic and mobility. However, all such efforts will only be accepted by the general public and the socio-political system, if the social achievements of increased mobility, which have contributed to the quality of life of the individual will not be abandoned.

Given the value system of the "post-modern" society, the fashionable life styles of the increasingly scattered target groups of the civilsociety, this is an extremely difficult task. And it would be inacceptable if mobility becomes again a class-related privilege, for which the growing underclass in Europe has to suffer.

With respect to the indispensable protection of the endangered environment and to the need for sustainable development in cities and regions of Europe, answers to fundamental transport and communication questions will have to be found. Such questions are:

- To which extent is the European economy, is economic growth, inextricably linked to a continuous increase of transport and communication? Is the wealth of the West-European society based on individual mobility? What are the social costs of unconstrained mobility?
- How can sector policies, encouraging or even promoting a further increase of individual and goods mobility be monitored and changed? How can the process of political bargaining in the transport/land use game be structured?
- How can negative implications of the fashionable "just-in-time" philosophy for urban and regional development be avoided? What are the ways out of that approach which simply shifts storage costs from the private to the public sector.
- Is there a new hierarchy of European transport networks emerging and what are the likely consequences for cities and

regions? What happens to regions which are in the shadow of accessibility to the new mega-system?

- What are the negative spatial and environmental impacts of regional transportation centres emerging all over Europe to accomodate the logistic requirements of rapidly expanding network economies?
- What are the impacts of growing airport cities on urban development? Will they change the spatial structure of cities and their regional hinterlands?
- Are there ways and means and policies which may contribute to constrain unlimited goods traffic? Will such measures be politically accepted? Or how can political acceptability be achieved?
- Is a revival of strong land use control mechanisms indispensable to stop the consumption of land for other uses? What will be the likely implications of necessarily changing agrarian policies on land markets in densely built-up regions.
- How will the transport needs of an continuously aging population in urban and rural areas be met or reduced? Is leisure traffic to be a major concern in the future?

Related to such questions are more general politico-administrative, organisational and financial aspects of managing the daily increasing complexity of the transportation and communication system in advanced societies. Three such questions are:

- Who is legitimised and willing to intervene in the transport and communication system at which level? (European, national, regional or local.) What are the forces, who are the actors openly or informally impeding such intervention?
- Can the declining role of the public sector be compensated by public-private partnerships, and if yes, where and under which conditions can this be done in the transportation and communication sector?
- What is the role of the media in the transport/land use game. How can new information and communication technologies be used to convey mobility constraining messages to decision-makers and opinion leaders?

Despite scattered efforts here and there in Europe, the knowledge of the forces and mechanisms of managing mobility and transport issues is still imcomplete. Truly interdisciplinary research in this field is scarce. Qualified transdisciplinary education and training at higher levels is far less developed than the importance of the field in the affluent Western society would suggest.

HOW CAN TRANSPORTATION AND COMMUNICATION ISSUES BEST BE ACCOMODATED IN PLANNING SCHOOLS?

Given the complex character of the above questions it is obvious that planning education in its present shape can only scratch the ground and just deal with the most central issues of transportation

1. Structural and organisational context:

Staff: 17 professors and around 50 full time lecturers and researchers assigned to 16 Fachgebiete (subject areas/chairs) or to the joint research centre (IRPUD). The chairs are:

Urban Design and Land Use Planning, Urban and Regional Planning, Regional and National Planning, Urban Engineering and Water Resource Planning, Transport Planning, Surveying and Land Management, Ecology and Landscape Planning, Urban and Landscape Design, Regional Economics, Public Finance and Budgeting, Industrial and Commercial Development Planning, Political Economy, Sociology, Urban and Regional Geography, Planning Law, Systems Theory and Systems Engineering, and European Planning.

2. Character and duration of the course:

The course is a combined undergraduate and postgraduate course (min. 9 semesters, average 5 to 6 years).
In addition there is a two years Ms.Sc. programme in Regional Planning and Development (SPRING) for students from developing countries in Africa and Asia.

3. Degrees:

Dipl. Ing. *(Raumplanung)*, Dr.Ing./Dr. rer.pol. *(Raumplanung)*.

4. Student numbers:

1100 (1991) There is an annual (limited) intake of 145 students. 25 are accepted for the SPRING-programme.

5. Job market:

The 1200 graduates so far (since the school has been founded in 1969) are working in public, semi-public and private institutions and firms dealing, for example, with traditional urban planning or regional policy, with economic promotion, housing or developing, with transportation or environmental planning. Planners from Dortmund are also employed among others, by research organisations, political parties and the media.

Figure 4 a. *Planning education at the Dortmund School of Planning I: context and character.*

1. Curriculum characteristics:

- Integrating social, political, legal, administrative, economic, ecological, architectural and engineering aspects of planning at various decision making levels;
- Combining both scientific and professional contents, by integrating analytical, design and action orientated elements of planning;
- Creating awareness and sensibility for social and environmental aspects of spatial development;
- Adding information mangement and communication skills to traditional knowledge based capabilities;

Within the rather broad and general curriculum the students are free to specialise in any planning related subject area (e.g. transportation/land use systems, environmental planning, regional policy). The integration of the various subject areas takes place in small groups centred around real world projects;

2. Major subject areas:

1. and 2. year:
- Social science for planning (12o teaching hours in class)
- Economics for planning (135)
- Urban and regional planning (75)
- Urban engineering (105)
- Planning law and land economics(120)
- Ecology for planning(60)
- Systems theory and techniques for planning (135)
- Projects (360)

3. and 4. year:
- Planning and decision making theory and techniques (120)
- Theories and modells of spatial development (120)
- Methods, procedures and instruments of planning (90)
- Urban design and conservation (60)
- Housing (60)
- Transportation (60)
- Water resource management, energy and waste management (60)
- Landscape planning (60)
- Systems theory and systems engineering (90)
- Projects (360)

5. year:
- Oral examinations and diploma thesis

Figure 4 b. *Planning education at the Dortmund School of Planning II: curriculum*

within its core curriculum. Very few planning schools in Europe are sufficiently staffed to cover transport and communication issues in their complexity. (The former Hochschule für Verkehrswesen in Dresden may have been an exception.) Although the rather large planning school at the University of Dortmund has its own chair of transport, the subject area is not a major concern of the school's curriculum (Figure 4).

Schools of Civil Engineering, offering specialised courses in transportation planning in turn, are usually forced to neglect all the linkages and interactions to social, environmental and other economic facets of the subject area or just not interested to cover such facets in their curricula.

Transport related courses having been offered during 1990 and 1991 were:

1. Mandatory courses *(lectures):*

- Raumerschließung durch Verkehrssysteme (3rd semester)
- Verkehrsplanung 1: Verkehrssysteme (4)
- Verkehrsplanung 2: Verkehrssysteme (4)
- Verkehrsplanung 3: Quantitative Methoden zur Prognose des Verkehrsgeschehens (8)
- Verkehrsplanung 4: Verkehrswirtschaft und Verkehrspolitik (7)
- Verkehrsplanung 5: Gestaltung und Bewertung von Verkehrsnetzen (8)

2. Optional courses *(seminars):*

- Verkehrsprobleme in Ländern der Dritten Welt
- Systemanalytische Verfahren im Verkehrswesen
- Siedlungsentwicklung und Verkehr
- Güterverkehr auf Straße und Schiene
- Stadtverkehr im Wandel- Anspruch und Wirklichkeit
- Modelle räumlicher Interaktion: Raum-Zeit-Labor
- Transport planning (for SPRING-students from developing countries)

3. Projects *(themes of optional one year studios handled by groups of 10 to 15 students):*

- Autofreie Innenstadt Dortmund
- Umweltverträgliche kommunale Verkehrsplanung in Waltrop
- Verkehrswende in Wuppertal-Heckinghausen
- Fußgängerfreundlicher Ortsteil Kurl
- Verkehrsplanung aus Frauensicht
- ÖPNV-Verkehrskonzept für die Stadt Witten
- Tempo-30 Zonen in Dortmund

Figure 4 c. Planning education at the Dortmund School of Planning III: for example transport

Their are many good (and bad) reasons for the resulting (and very much regrettable) evidence that the (the non-technical) transport and communication issue seems to slightly fall in between the various spheres of higher education in Europe. Although it is one of the most crucial areas of future urban and regional development and policy affecting sustainable development and individual quality of life, it receives very little promotion. This may be due to the immobile traditions of sector

1. **Institutions** of higher education in Germany offering transportation planning as a formalised option within civil engineering education (selection):

- RWTH **Aachen**: Studienrichtung Verkehrwesen und Raumplanung *(transportation and planning)*
- Technische Universität **Berlin**: Studienrichtung Verkehrs- und Raumplanung *(transportation and planning)*
- Universität **Bochum**: Vertiefungsrichtung Verkehrswesen *(transportation)*
- Universität **Karlsruhe**: Vertiefungsrichtung Verkehrswesen und Städtebau *(transportation and urban planning)*
- Technische Universität **Hannover**: Vertiefungsrichtung Verkehrswesen *(transportation)*
- Universität der Bundeswehr **München**: Studienrichtung Verkehrswesen und Raumplanung *(transportation and planning)*
- Gesamthochschule **Wuppertal**: Studienrichtung Verkehrsplanung und Verkehrssysteme *(transport planning and transport systems)*

Transportation as a special course option is also offerered at the Fachhochschulen in Aachen, Stuttgart, Münster or Siegen.

2. **Example:** urban and regional planning courses being taught to civil engineering students at the University of Karlsruhe in 1991 were:

- Landesplanung *(regional planning at state level)*
- Städtebau *(urban planning)*
- Öffentlicher Haushalt *(public finance)*
- Raumordnungspolitik *(national physical planning policy)*
- Kolloquium für Städtebau und Landesplanung *(urban and regional planning colloquium)*

Sources: Bundesanstalt für Arbeit (ed.), Einrichtungen zur beruflichen Bildung, Nürnberg 1991; and Vorlesungsver zeichnisse (university directories).

Figure 5. Transportation planning education in Germany.

disciplines and their mutual jealousy, to the lack of incentives for interdisciplinary studies and research, the departmental structure of research funding agencies with their traditional clientele not beeing keen to remove the boundaries and open the claims, or to the power and influence of the corporate industries profiting from growing mobility. Their research divisions are usually much more advanced than the few scattered research units at universities. It seems that the universities are seriously lagging behind. However, what is needed to overcome the shortcomings, sketched above?

- *Obviously, More Interdisciplinary Research:* what is needed to alter present conditions is above all a strengthening of research in transport and communication related issues of urban and regional development. Given the constraints sketched above, only through international co-operation existing institutional and psychological barriers may be removed. First studies commissioned by the European Commission in preparing for the Europe 2000 document and efforts of the European Science Foundation, may be a step in that direction, although they cannot compensate for the lack of a transnational research scheme (and funds!) for basic research in the subject area. Academic education as well as professional training in turn are based on knowledge acquired from doing basic and applied research.

- *Internationalisation of Planning Education:* There are many reasons why planning education must become more international. There is the continuous internationalisation of the economy which requires a much wider view of factors influencing spatial development in cities and regions. There is the increasing international competition among cities and regions, which forces local and regional planners to look outside their day-to-day environment if they don't want to become the mere playmates of international investors and developers, or just wish to attract floating foot-loose capital from Japan or the Gulf Emirates. The establishment of a transnational tier of government at the European level is another reason of widening the perspectives of planning education, as more and more decisions are taken at this level, changing economic or social conditions of development or setting new transnational norms and standards for local or regional development. The evolution of a European wide job market for planners may be a reason which urges planning education to internationalise their curricula. Since the beginning of the nineties, international developers are searching for highly qualified, multilingual planners ready to move around the Europe. They compete with larger research institutions and consultancy firms for the talented and mobile European planners.

There are various ways and different approaches to internationalise planning education. To begin with the internationalisation of planning education has to be seen in the wider context of the slowly converging system of higher education in Europe. Over long, and in some countries earlier than in others, the system of higher education will be characterised by four levels of education after various forms of basic education which qualify for entering the higher education system: These levels are

- undergraduate education: usually a three to four years rather general and basic course in practice oriented planning, laying broad foundations for a professional career in the respective region or country.
- graduate education: a two years course focusing on theories and policies of planning both urban and regional.
- specialised education: a one-to-four years course either offering specialised professional education in a selected problem area (e.g. transport and communication, urban design, environmental planning, national planning or planning in developing countries) or more academically oriented doctoral studies.
- continuous education: short courses of usually one day to six weeks duration aiming at either refreshing and updating active knowledge or acquiring new professional knowledge in a certain "window" of the discipline.

This system of higher education (Figure 6) will be the future organisational framework for European planning schools and for their efforts to meet the demand for planning education in the respective region or country.

Obviously, international aspects will have to be considered out all four levels, although in different ways. The fashionable slogan "Think globally, act locally" may describe the general direction, but the local outline and design of the programme will differ from place to place. It will depend on resources staff interests and language capabilities of the respective place.

Depending on such conditions and on the willingness, the capabilities and influence and the bargaining power of local institutions and actors to alter or to expand the established system within its university system. The major ways and strands of possible action are:

- *Promoting international student and staff exchange:* Since a few years, European student excange programmes, such as

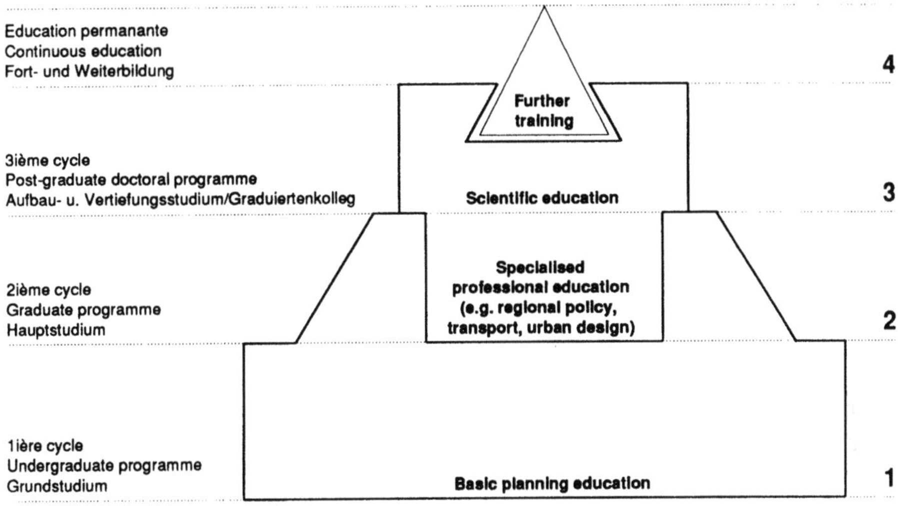

Figure 6. *The convergent system of planning education in Europe.*

ERASMUS and TEMPUS, have very much contributed to a greater mobility of students and staff. The number of joint graduate and post-graduate programmes among universities and other institutions of higher education in Europe has considerably increased over the last decade. This has been pioneered by business schools and departments of law and economics. Many planning schools in Britain, Germany and France have encouraged their students to go abroad for a limited time period to get an insight in other planning systems and an outside perspective of their own system, but also to train their language skills. However, this can only be a start. Much more has still to be done to institutionalise and to promote the exchange of students, to convince national institutions concerned to increase their funds for such activities, and to meet the raising interest among students to go abroad. The planning schools themselves may be have to scrutinize their routine teaching, whether it is suitable for foreign students, and eventually add substantial and tutorial elements to their curricula tailored to their needs and expectations.

The public and private transport and communication sector would additionally be an appropriate field for promoting foreign internships, another instrument to widen the international horizon of planners during their university education.

- *Internationalising the core curriculum:* Given the increasing internatinalisation of local and regional economies, planning education has to consider the resulting consequences for urban and regional development. This has various implications for planning courses who aim at competing in the international market. First, the knowledge of international capital flows and investment patterns and their likely consequences for urban and regional development, as well as a profound knowledge of theories and strategies of international economic and spatial development becomes more and more indispensable for planners, even for those who are not aiming at an international professional career. This is essential to understand the international forces affecting local decisions on spatial development. Second, more information must be disseminated to planning students on such international (mainly European) institutions and instruments which either constrain or promote local economic and spatial development, institutions which in an economically united Europe assemble and execute more and more political power.

Obviously, transportation and communication related aspects of spatial development ought to be essential elements in such a core curriculum, primarily those of economic, social and behavioural nature.

A pragmatic proposal for a European core curriculum covering the above mentioned contents has recently been made by a working group of the Association of European Schools of Planning (see Figure 7).

1. Reasons and objectives:

- to better understand the cultural, socio-economic and politico-administrative context of urban and regional planning in Europe;
- to identify and assess the effects of international economic and geo-political development trends on cities and regions;
- to better understand the mechanisms of the own planning system;
- to be able to communicate with planners from other countries.

2. Suggested framework and contents:

Modul 1: **Planning in Europe:** urban and regional problems, urban systems and hierachies, socio-economic and politico-administrative context, planning systems and instruments.

Modul 2: **Planning for Europe:** activities, policies and strategies of European institutions (EC, OECD, Council of Europe, ECE).

Modul 3: **Planning in another European country** (exemplary): planning traditions, problems and perspectives, institutions and actors, statutory framework, urban, regional and sectoral policies.

3. Structural preconditions:

- qualified staff has foreign language capabilities;
- teaching is based on and linked to researchprojects;
- sufficient budgets for study tours are available;
- libraries are equipped with foreign language literature.

Source: after Albrechts, L. et.al., (ed.), Towards a European Core Curriculum in Planning Education. AESOP-Papers No. 6, Loewen/Dortmund/Newcastle/Grenoble, 1990.

Figure 7. The European Core Curriculum.

Research Cooperation Among European Planning Schools!

Unless individual persons or university research units endeavour to initiate comparative studies and successfully acquire the necessary funds for such efforts, research cooperation among scientists from planning schools is still the exception. Only recently the European Community has commissioned a number of applied research projects to joint ventures in the planning sector. This initiative, started by the need to get up-to-date information for the Europe 2000 document, has demonstrated that such research cooperation is possible and fruitful, if (financial and image) incentives are available. It has, however, also revealed that only few planning schools at universities are really in a position to submit proposals and to enter the competition, leaving the way to more flexible (and more efficient?) consulting firms.

Given the complexity of the transportation and communication sector it may be indispensable for planning schools to join the intellectual and managerial capabilities and forces of their senior staff for transnational research ventures in that field.

Division of Labour Among Post-graduate Schools in Europe

One could argue that any division of labour per se contributes to an increasing mobility and hence, has to be avoided. Given the small number of planning schools in their respective countries, however, only a specialisation and concentration of power could create the structural preconditions to assemble the necessary intellectual power and strength. Following such lines of thinking, a European post-graduate or doctoral course for planners, specialising in transportation and communication aspects of spatial development could be offered for example by a Swedish institute of higher education, whereas a tourism specialisation could be installed at a Spanish university. Or an in – depth regional policy programme in Germany. Such a division of labour within Europe would offer new incentives for staff and students and it would attract international research funds to the respective school, where research, education and training in the selected field of studies are necessarily highly interlinkes.

Establishing Joint European Planning Schools

Another approach to overcome the problem of (occasionally far too) small numbers of staff and students in planning schools, and problems caused by small student numbers in a specialised field of studies, is to establish joint planning schools, such as the Nordisk Institute of Samhällsplanering. Establishing such international schools requires considerable organisational and political efforts and strength to overcome all the administrative and mental barrieres hindering the support of crossborder institutions and activities. However, in a joint Europe there are some chances that such difficulties may diminuish. Language will remain a constraint. But whether we like it or not, English will probably become the lingua franca of academia in Europe, at least in North and Northwest Europe. Additional there are good chances that new individual translation technologies will reach a new degree of sophistication and allow individuals to easily translate documents from one language into another one.

There are some good arguments to combine this approach with the proposal to offer specialised academic courses made earlier. Under certain conditions it could facilitate the institutionalisation and implementation process of innovative institutions.

It would be a misinterpretation of the quest for internationalising planning education if, by such a move, local and regional ground would be lost. Obviously, it depends very much on the level of higher education which is under discussion. Taking off from the endogenous regional or national socio-cultural and politico-administrative ground would certainly additionally weaken the few chances to really change things, to implement policies locally or regionally. Given the linkage between theory and practice in this academic discipline it may not even be leading to better theories of urban or regional development. Internationalising the ivory tower does not per se change its nature.

OUTLOOK

The changing geo-political environment in Europe and the resulting economic, social and political implications leave little room (and financial means) for necessary innovative reforms of the higher education system in Western and Northern Europe. This may be deplored. It limits the opportunities to any kind of experiment in higher education. Hence incrementalism and individual efforts will determine the agenda. Planning education, independant from traditional disciplines is, for various reasons, not yet fully accepted all over Europe at least not in traditional academic environments.

The deliberate and continuous misinterpretation of urban and regional planning as a top-down public sector control approach or as a fancy design discipline constraining individual freedom of architectural expression does not facilitate the process of innovation within a discipline. It rather hinders self-criticism and explorative innovations, which in any academic discipline are an indispensable prerequisite of development. Setting on and strengthening the international dimension of planning education is most probably, the only way out of the cul-de-sac, in which the discipline is occasionally driven to defend its existance against jealous departmental fathers and brothers. The complex transportation and communication element of the discipline is certainly one of the most convincing subject areas to demonstrate the future orientated character of the discipline. It is still underrated and insufficiently developed and it waits for initiatives and joint efforts.

PART IV:

ENVIRONMENT PERSPECTIVES

THE ENVIRONMENTAL SITUATION IN EUROPE

Lennart Schotte
Royal Swedish Academy of Agriculture and Forestry
Sweden

Among all the environmental problems, air pollution has been recognized as the most serious threat. Lakes and soils have been acidified. Forests are damaged in vast areas. We are dealing with chemical stress factors which in complicated mechanisms are causing disturbances in the ecosystem. There is widespread concern within the scientific, industrial, labour, economic, regulatory and public sectors of European society, that a continuation of recent trends in forest decline may lead to a number of undesirable consequences.

The scope of the problem is international. The scale however, of each instance of decline is local to regional. The spectrum of air pollutants differs from region to region. Thus while strategies to combat undesirable impacts must be related appropriately to the local and regional scales. The strategies must be set firmly in the context of the international arena to take account of transboundary air pollution and trade in raw materials for processing.

In order to reduce the impact on the environment, the Swedish forest industry invest around 20 percent of its total to protect and improve our environment.

We depend on harmony in our ecosystems, and emissions and pollutions are serious threats. The trend shows that they will increase in the future.

Another important part of the environmental pattern is the emission of heavy metals. If we try to summarize the situation and look upon year 2000 as the target one will find sulphur dioxide and nitrogen oxide emission patterns in different regions of Europe (Table 1).

Table 1. Emissions of sulphur and nitrogen expressed in kt

	SO_2			NO_x (as NO_2) 1)		
	1980	2000		1980	2000	
Nordic	1205	675	(1.9%)	751	526	(3.4%)
EEC-9	17048	10020	(28.8%)	10139	7916	(51.2%)
Central	481	170	(0.5%)	356	250	(1.6%)
Southern	6172	9873	(28.3%)	2018	2936	(19.0%)
Eastern	16332	14092	(40.5%)	3631	3841	(24.8%)
Europe	41238	34830		16895	15469	
	- 18%			- 9%		

1) Quantitative estimates are only available for NO_x emissions. If NH_3 emissions are taken into account the total Nitrogen emissions will be about double of the figures presented for NO_x.

We have no reason, unfortunately, to expect an improvement of more than 18 percent for nitrogen oxide. You will also find the distribution of improvements in different areas.

Another aspect of importance is the crossborder emission. Like most ther countries. Sweden is receiving quite a lot of pullutions from a number of neighboring countries. However, we also export some. The net effect is that we receive some 80-85 percent from abroad.

Table 2 shows the amount of sulphur dioxide emissions in Sweden, and Table 3 show the negative effects of road traffic.

There are lots of patterns that illustrats the environmental situation. In 1951 a powerline was built in the Southwest part of Sweden. In order to decide what material to use for the poles, the total area of the route was examined. The same area, was checked some 36 years later. The difference, the decrease in pH value by two units during the period is a pure effect of the deposition in this part of Sweden. The soil is a hundred times more acidic. The only explanation is the deposition of acidifying chemicals, mainly sulphur and nitrogen oxides.

Table 2. *Sulphur dioxide emission in Sweden*

Current emissions and anticipated emissions, under present environmental legislation (including the tighter requirements which will apply from 1993). Thousands of tonnes.

	1980	1988	1995	2000	2015	
					Low	High
Transport	35	33	38	40	45	50
Combustion						
- Oil	318	70	60	60	55	70
- Coal	6	19	24	24	35	62
- Misc.	4	8	8	8	5	8
Industrial processes	151	82	50	50	50	50
Total	514	213	180	182	190	240

Source: Swedish Environment Protection Board

Table 3. *Nitrogen emission in Sweden*

Current emissions and anticipated changes under present legislation. Thousands of tonnes.

	1980	1988	1995	2000	2015
Road traffic					
- Light vehicles	106	117	63	58	57
- Heavy vehicles	56	75	69	62	58
Shipping	23	33	40	45	48
Air traffic	6	7	10	13	15
Work vehicles	67	72	72	73	73
Railways	1	1	1	1	1
Combustion					
- Oil/gas	86	32	39	39	49
- Coal	3	17	12	12	45
- Misc.	7	15	12	12	17
Industrial proc.	34	17	15	15	15
Total	389	386	333	330	378

If we return to forest, there is always the discussion of what kind of contribution to the total stress factors air pollution represent. There are always natural stress in forest ecosystems, hitherto now we can observe new types of stress factors, caused by air pollution constituting a threat to, and a damage of the trees. In Europe there is always this discussion whether the nitrogen emissions and precipitations cause damages to the forests by fertilization.

However the precipitation of nitrogenoxides in Sweden is contributed to by the road traffic in Sweden and its neighboring countries.

It is true that nitrogen being added to the forest ecosystem means a fertilizing, means improvement in their growth. The Northern part of Sweden differs quite a lot from the Southern part. But, the truth is more complicated. The growth here, the fertilizing effect is masking the negative effect in the ecosystems. It goes to a certain reversal point, where the ecosystem receives too much of nitrogen. That means again that with nitrate ions, nutrients are being removed from the soil and the system, thereby making the situation less favorable for the trees. We will see a decrease in the growth-rate which is important. Too many speak about forest-death and compare the situation to the utmost serious one down in the mountain areas of former DDR, Czechoslovakia and Poland. But the true risk, the true threat to the future in most areas is rather the lingering effect of lower growth-rate in years to come, and irreversible change in the ecosystems.

Different parts of Europe differ in sensitivity. The Nordic countries are in a specific situation and ought to prevent future effects.

There is no doubt, there are depositions from the emissions which are serious for the future. With such a background it is adequate to ask yourself, what kind of depositions may be tolerated? Then the concept of critical loads, limits of survival on a sustainable basis for sensitive species in different ecosystems is a good one.

I will in short give you some ideas on this. This is again related to Sweden. In the country as a whole, of course, the weathering capacity is of importance. This shows the relation between sulphur deposition and weathering (Table 4). With a very slow rate of weathering the ecosystem can tolerate some 3 kilogram per hectar and year and with a very rapid weathering up to 30-35.

Table 4. *Critical loads for acid and sulphur in relation to the weathering capacity of the soil in forest land.*

Class	Acid input (KEQ H+/KM^2/YR)	Sulphur deposition (KG S/HA/YR)
1. Very slow weathering	< 20	< 3
2. Slow weathering	20 - 50	3 - 8
3. Moderate weathering	50 - 100	8 - 16
4. Rapid weathering	100 - 200	16 - 32
5. Very rapid weathering	> 200	> 32

Another component in this discussion is ozone, being formed catalyticly from nitrogen oxides and hydro carbons, thereby related to traffic. Considering the critical levels of ozone (Table 5) and the pattern of emissions and depositions, it is necessary to question the amount of excess deposition of sulphur above this target.

Table 5. *Critical levels of ozone for sensitive plants, plant communities, and ecosystems.*

Exposure (hours)	Ozone concentration (UG/m^3)
0.5	300
1.0	150
2.0	110
4.0	80
8.0	60
Vegetation period*	50

* Daily mean value during the summer half-year.

There are some discussions going on the moderate the critical load system, to a target load where technical methods and the economic development comes into the picture. Having done that, thereby slightly moderating the concept of critical load, we will still see large areas receiving much more than acceptable.

The International Institute of Applied System Analysis (IIASA) in Luxemburg outside Vienna has made quite a number of studies on the ecosystem and the forest situation (Table 6).

Table 6. *Extent of forests exceeding the critical loads for sulphur and nitrogen. For ozone, number of times exceeding the actual critical load*

	Sulphur Coniferous		Deciduous		Nitrogen Coniferous	Deciduous	Ozone
	1985	2000	1985	2000	1985-2000	1985-2000	1985-2000
Nordic	59%	48%	19%	7%	75%	52%	~2 times
EEC-9	88%	76%	34%	24%	83%	55%	~2 times
Central	98%	93%	50%	46%	100%	86%	~2.5 times
Southern	62%	84%	18%	40%	34%	21%	n.a
Eastern	98%	98%	84%	76%	76%	47%	~2.5 times

This gives us a warning for the future even if we are very actively dealing internationally with negotiations to reduce the emissions!

In some areas of Europe, including parts of Sweden, we have a stress overloaded system. If you drive a car without a catalytic-converter through these areas, you are causing a marginal effect upon the ecosystem. I have therefore proposed in the Environmental Council of the Swedish National Road Administration that there should be some signs put up by the side of the road, just telling the driver and the passenger that they are passing into an area which is specifically sensitive to margin effects. It should be, maybe temporar limitation of speed in that area, as long as the proportion of cars with catalytic-converters is small, and as long as the pollutant depositions from different sources, are at present level. When sufficient improvement has been established, the restrictions should be eliminated. This might be incentives for the traffic road users to see that he or she should be a little more careful and those individual actions have significant and immediate effects.

A Swedish Company with a license of a Swiss patent, produces and offers to the market of the Nordic countries a diesel-cleaning system. Water to an extent of some 15 to 20 percent is inserted into the diesel-oil, forming particles with a core of water which are injected into the combustion chamber, where they explode and form very small drops. These drops can be effectedly burned at lower burning temperature with the larger area giving a higher efficiency. By this, the strain on the catalytic-converter is lowered so you can combine the diesel-cleaning system, quite easy, with the converter.

With present pollution rates we have to accept, or at least see the threat, of a reduction of some 85 million m^3 per year in growth of European forests (Table 7), with the exception of the European parts of the USSR. In total 15-20 percent of the present growth in fact sustainably negative effects of the environmental loading in Europe today.

Table 7. *Decrease in million m^3 per year in the potential wood supply during the next 100 years.*

Nordic	11.0 million
EEC-9	23.9 million
Central	5.8 million
Southern	10.2 million
Eastern	34.4 million

The environmental problems of our time are not only global by nature and insensitive to national borders and ideological barriers. They are as much transsectorial as they are transboundary or cross continental.

There may still be those who say that there is not enough proof to show that the acid emissions from coal and oil-fired energy plants, industrial activities, traffic and agriculture constitute a significant cause to the death of European forests. If we demand full scientific proof of each and every one of all the aspects of the environmental decline before acting, nothing will happen until it is to late. Also there will always be those who deny the environmental risks in their field. They may argue in the name of scientific prudence even when their real reason is short time economic concerns. We should not forget, that what is happening to our environment has a lot to do with security. Let us also remember, that the complete control of the air pollution problem really is in hands of mankind and no doubt, I am convinced, possible to achieve. The limitations are time and financing. The solution demands however a true international cooperation!

TRANSPORTATION AND THE ENVIRONMENT

Kenneth A. Small
Harvard University and University of California, Irvine
U.S.A.

It is widely believed that environmental effects must increasingly play a dominant role in shaping the transportation systems of the future. If true, this belief has important implications because it suggests a departure from traditional means of evaluating transportation policies. Indeed, this type of argument has been used to justify projects that fail to meet traditional criteria, and furthermore to predict significant changes in personal mobility patterns as they have been shaped by the automobile era. Therefore, it is of interest to examine whether and in what form the belief is true.

Specifically, what is the most likely type of adjustment that our societies will make to diminish the potential adverse environmental impacts of transportation activities? Will it take the form mainly of new technology, or will it involve alterations in individual behavior? That is to say, will the engineers be developing the solutions, or will we as ordinary citizens have to take an active role? The answer makes a great deal of difference not only to the way that governments and administrators must approach these problems, but to the kinds of political action that will be required to deal with environmental impacts.

In addressing this question, I pretend to no special futuristic vision. I am not associated with the Worldwatch Institute; I routinely avoid predicting vast social trends; I even avoid predicting next month's interest rates or stock market prices. However, we know something about many of the relevant factors, and it is enlightening to think through their implications. Indeed, many scenarios can be eliminated just by making some quantitative observations about the costs and behavioral propensities that help shape people's responses to transportation and environmental problems.

I focus mainly on what *might* happen, not what *should* happen. To this end, I explore the broad implications of foreseeable trends, in conjunction with our best current knowledge of technologies and costs. Still, it is necessary to make some normative assumption about the nature of policy response, for otherwise we can predict nothing whatever. The reason is that *sufficiently bad public policy can make almost anything happen*. For example, a program costing $50 to $150 billion dollars can be mandated to remove asbestos fibers from buildings in the United States, even though asbestos is not released into the air except during structural modifications (such as in asbestos removal), and even though 95 percent of asbestos in U.S. buildings is of a type (chrysotile) not implicated in lung cancer (Abelson, 1990).

Since truly perverse public policies are very difficult to predict, I will limit myself to an analysis which assumes that reasonably rational policies prevail. To make this notion more concrete, let us say that current policies, for all their defects, will not be replaced by policies that are substantially worse from the point of view of long-run economic efficiency. This rather ambiguous assumption leaves a good deal of room for judgment; but I will make the case that it narrows considerably the range of policies toward environmental effects that otherwise might be thought likely.

SOME RELEVANT TRENDS

As background to our inquiry, let us note four major trends affecting the environmental consequences of transportation activities.

First is the remarkable and continuing rise in ownership and use of automobiles. Although the United States is most notable for its high *levels* of automobile use, the upward *trends* are universal in the developed world. For example, between 1970 and 1987, vehicle-kilometers traveled by automobiles rose 66 percent in the U.S. and 72 percent in eleven nations of Western Europe for which data are available. Within the latter group is a range of 17-year growth rates from 41 percent in Denmark to 181 percent in Spain.

(Sweden's rate was 54 percent.)[1] The trends in both continents indicate that even where transit service is very convenient, people prefer the greater flexibility of a personal vehicle when they can afford it, and increasingly they can afford it. Whether these trends are caused mainly by inexorable events or by public policies, possibly faulty ones, they are deeply imbedded in current economic and social systems and will not be reversed easily.

Second, and less known, is an apparent tendency for automobile ownership to level off when it approaches one personal vehicle per adult, as it is doing in the U.S. This is well documented by Lave (1990), who points to several regions in the U.S. where the ratio of light-duty passenger vehicles to driving-age population grew rapidly but slowed as it approached a value of 1.0. It is less certain whether this will cause per-capita usage of these vehicles to level off also; but it seems likely because distance traveled per automobile has been quite constant in the U.S., and total time spent traveling per person has shown remarkable regularities across both nations and time periods. Only if further road building brings about an increase in average speeds is the annual distance traveled by each automobile likely to rise very much.

The third trend is the dramatic reduction in noxious air-pollutant emissions and energy consumption per vehicle-kilometer travelled. In the case of noxious emissions, this is primarily a result of government regulation. In the U.S., emissions of hydrocarbons, carbon monoxide, and nitrogen oxides from new cars had been reduced respectively by 87 percent, 85 percent, and 62 percent by the late 1980s; and stringent new requirements have been legislated for the 1990s. Comparable control levels are in effect in Austria, Norway, Sweden, Switzerland, Australia, Canada, and South Korea.[2] In the case of energy consumption, the reduction is smaller and has resulted from a combination of taxation policies (especially in Europe), direct regulation (especially in the U.S.), and modest pressures from two sharp rises in energy prices between

---

1 European figures are computed from: European Conference of Ministers of Transport, **Statistical Trends in Transport 1965-1987**, Table 3-1-6, p. 124. The eleven nations are all those for which the statistics are given except Finland (for which there appears to be an error), viz.: Austria, Belgium, Denmark, France, Germany, Italy, Netherlands, Norway, Spain, Sweden, and U.K. U.S. figures are computed from U.S. Federal Highway Administation, **Highway Statistics**, various years; they include passenger automobiles plus 57 percent of two-axle four-tire trucks, that being the percentage found in surveys to be used primarily for passenger transportation according to Lave (1990). The U.S. growth rate is even less if no trucks are counted.

2 Renner (1988), pp. 3940.

1972 and 1980. Of course, average fuel consumption for all autos in use responds more slowly than that for new cars, but even that measure declined about 30 percent in the U.S. between 1970 and 1987.[3]

The fourth trend is the increasing difficulty experienced by government bodies charged with financing transportation infrastructure. This financial squeeze is often associated with inability to prevent severe road congestion in crowded or rapidly growing urban areas, where new construction can be very expensive; but other types of infrastructure are also affected, suggesting that broader financial trends are at work.[4] In Europe, economic integration can only exacerbate this problem as industries consolidate and international trade expands. The problem of financing roads has led to considerable interest in toll roads and private-sector financing, especially in the U.S. and U.K.

ENVIRONMENTAL EFFECTS OF TRANSPORTATION: SOME MAGNITUDES

Of the many environmental issues associated with transportation, it is useful to highlight three which seem likely to have the greatest influence on the development of transportation systems over the next few decades. First in importance to policy up to now is air pollution by noxious gases. A second influence, just emerging but potentially enormous, is the contribution of automotive emissions to greenhouse warming, a poorly understood but possibly dramatic effect of certain gases on the earth's temperature. A third influence is traffic congestion, which might not be classified as an environmental effect but which has demonstrable adverse psychological effects on health and certainly ranks as one of the greatest perceived problems with our current transportation system.

Other problems, such as noise, land impacts, and aesthetics, may also be quite important but do not seem to have such urgent impacts on thinking about public policy. They are therefore not considered here.

---

[3] Calculated from Davis et.al. (1989), Table 2.14, p. 2-25.

[4] See, for example, U.S. National Council on Public Works Improvement (1988) or U.S. Congressional Budget Office (1988).

Air Pollution

The chief air pollutants resulting from automobile use are carbon monoxide, nitrogen oxides (both as direct irritants and as contributors to acid rain), and ozone (produced by emissions of hydrocarbons and nitrogen oxides). In addition, diesel engines produce small airborne particles and sulfur oxides, two of the most well documented health hazards in ambient air.

It is helpful to have some notion of the magnitude of the health and other costs of these pollutants. Estimating them is a very uncertain enterprise, but the principle is fairly clear: we wish to know how much people would be willing to pay to fully mitigate the problems – for example, through voting on some national health program, or on a plan to nullify the effect of acid rain on lakes and forests. (This presumes they would be well informed and vote their true wishes.) Such assessments may be made through questionnaires or, less directly, through observations of people's behaviour in the face of various health- or environment-related personal decisions such as job choice.

For the United States, estimates of costs from the pollution caused by automobiles are on the order of $20-40 billion per year.[5] This translates into about $1000-$2000 over the life of each car.[6] If these estimates are of the correct order of magnitude, they support two conclusions: first, that pollution costs are large, a substantial

---

5 This is my assessment of the most reasonable range for damages resulting from all emissions of carbon monoxide, hydrocarbons, and nitrogen oxides (including seconary products such as ozone), which may be taken as roughly representing the problems caused by automobiles. Of course other sources contribute to these emissions, expecially to nitrogen oxides, but automobiles are dominant; contrariwise automobiles emit other pollutants, but account for only a small fraction of them. The figures given are roughly the central values found in a variety of often conflicting studies. For example, Small (1977, p. 121) estimates 1970 U.S. costs from these pollutants at $5.38 billion, but uses a value of life substantially lower than current estimates based on labor-market behavior toward risk (Small, 1991, p. 83). U.S. Federal Highway Administration (1982, p. E-47) uses estimates, derived from Haugaard (1981), that add to $25 billion for 1981. The Los Angeles region, where ozone is the worst automotive-related air-quality problem, typically accounts for a substantial fraction of such estimates; a recent estimate of the health costs of exceeding the federal ozone standards in the Los Angeles region is $2.7 billion (Brajer et.al., 1991), a figure believed too high by Krupnick and Portney (1991).

6 Based on the approximation of 200 million cars, each lasting 10 years. Cameron (1991, p. 21), using a rather generous allocation of responsibility for health effects to automobiles, estimates their costs in Southern California at $0.06 per mile or about $6,000 over the life of a car.

fraction of the purchase price of a vehicle; and second, that pollution costs are not nearly so large as the total costs of driving, especially if we include the value of people's time.

These conclusions have two important implications. First, considerable up-front expense can be justified in the form of pollution-control technology to substantially reduce emissions. (This is not to take a position either way on the controversy over whether current regulations are justified.)[7] Second, people have already demonstrated their willingness to incur money and time costs for automobile travel that far exceed the quantifiable damages from air pollution. To put it differently, if people were forced to choose between cleaner air and the mobility afforded by personal vehicles, it appears that they would choose mobility. Fortunately, advances in pollution-control technology have so far made this choice unnecessary, and will probably continue to do so.

Greenhouse Effect

Worldwide industrialization has begun to load the atmosphere with gases that are mostly quite harmless to breathing humans, but that are subtly raising the earth's temperature by trapping reflected heat from its surface in a process known as the "greenhouse effect". The most significant such "greenhouse gases" are carbon dioxide (CO_2, exhaled by people, cars, power plants, and anything else that uses oxygen to burn up fuel) and clorfluoro-hydrocarbons (CFCs, used in many products including refrigerators and air conditioners).

CFCs, while small in total quantity, are so powerful that a car with a CFC-charged air conditioner is believed to contribute more to the greenhouse effect from loss of its refrigerant than from all the carbon dioxide emitted throughout the car's lifetime. In the U.S., automobiles account for about one-fourth of all CFC emissions.[8] However, CFCs are being phased out for a more urgent reason, namely their effect on stratospheric ozone and thereby on ultraviolet radiation reaching the earth. As a result, the main greenhouse gas of concern to future transportation policy is probably CO_2, the inevitable combustion product of any carbon-based fuel such as petroleum, natural gas, or coal.

---

7 See, for example, Crandall et.al. (1986).,

8 Gordon (1991), pp. 57-69.

The precise environmental effects of greenhouse gases are extremely uncertain, both in direction and magnitude. It is not even certain that they are harmful on balance, since many agricultural regions could benefit from warmer temperatures and from higher concentrations of atmospheric CO_2.[9] Furthermore, some scientists have expressed hope that measures can be found to offset the greenhouse effect by deliberately cooling the earth, for example by placing small particles in the stratosphere or by promoting photosynthesis in ocean algae (which would remove carbon dioxide from the air). It is therefore premature to assert what policy stance will ultimately prove to be sound. Nevertheless, enough evidence has accumulated to make immediate CO_2 reduction at least a "reasonable" policy, pending further scientific evidence. International negotiations to bring this about have already made considerable headway.

Nordhaus (1991a) has reviewed estimates of the cost of various degrees of CO_2 reduction. Figure 1 portrays his results, in the form of a curve relating the marginal (incremental) cost of CO_2 reduction to the extent of that reduction. Each point on the curve represents a mix of policies, judged to be the most efficient possible to achieve the amount of reduction indicated on the horizontal axis. The height of the curve represents the cost, per kilogram of reduced emission of elemental carbon, of tightening policies slightly further.

A rather drastic policy, probably far more than will be agreed upon in the forseeable future, would be to reduce emissions by 50 percent from what they would otherwise be. Nordhaus's analysis, as portrayed in Figure 1, suggests that the efficient package of policies to achieve this would have a marginal cost of around \$0.13 per kilogram of carbon removed. This means that the transportation policies that would be part of that package would, either by regulation or taxation, eliminate activities whose value to users does not exceed their current costs by \$0.13 for every kilogram of carbon they produce. That is, they would have effects comparable to those of a tax of \$0.13 /kg on the carbon content of fuels. Such a tax, with maximum rate about two-thirds this amount, was in fact adopted, but subsequently withdrawn, by the European Community.[10]

---

[9] The most comprehensive assessment is IPCC (1990). A very readable account of the scientific evidence is given by Greenaway (1991). Economic effects are analyzed cogently by Nordhaus (1991b).

[10] **Wall Street Journal**, "EC Brakes Energy Proposals", 30 October, 1991, p. A12. This level of carbon tax is also consistent with the analysis of Jorgenson and

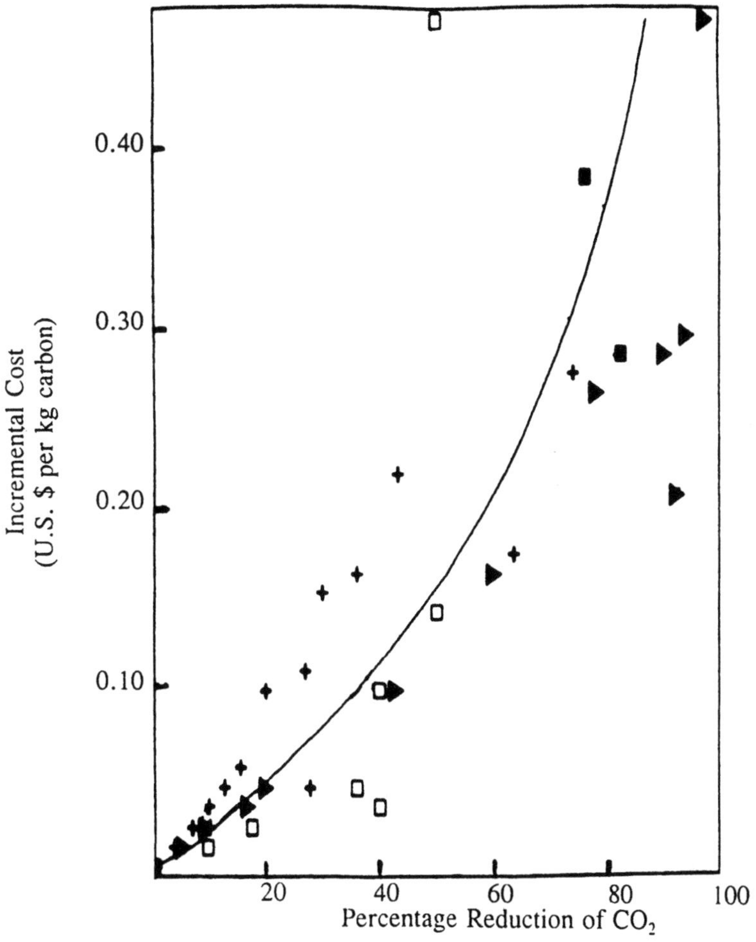

Figure 1. ***Marginal cost of carbon-dioxide reduction.***

Adapted from: William Nordhaus, The Energy Journal, 12 (1), January 1991, copyright 1991 by the International Association for Energy Economics.

Traffic Congestion

The costs of traffic congestion have been estimated in a variety of ways. A recent estimate for the Los Angeles region was \$1.8 billion in 1984.[11] It e such large numbers: a single commuter delayed by 15

Wilcoxen (1991, p. 24), who estimate that a carbon tax of \$0.066/kg would reduce U.S. CO_2 emissions by 32 percent below base case in the year 2020.

11 SCAG (1988).

minutes each way on a work trip will typically value this delay at around $2.40 per day, based on an average value of time of approximately $4.80 per hour derived from studies in the United States, Great Britain, and Australia.[12] With a typical work schedule of 250 days per year, such a commuter incurs costs of $600 per year due to the delays. If the 791,000 workers in the four largest employment centers in Los Angeles[13] each suffer this cost, we are already in the range of $475 million per year. Many trips to other work areas, and many nonwork trips, also suffer severe congestion.

On a per-vehicle basis, these costs are considerably higher than those estimated for noxious air pollution and the greenhouse effect in the previous subsections. Added to this is the tension, uncertainty in travel schedules, and frustration at a highway system that just doesn't seem to be working the way it is supposed to. So long as delays of this magnitude continue to typify peak-hour travel conditions, congestion can be expected to remain a top concern in transportation policy.

SOLUTIONS TO ENVIRONMENTAL PROBLEMS

We now consider the implications of these efforts to quantify the magnitudes of environmental effects. The approach is to ask what travelers or regulators are likely to do in response to each problem, given the magnitudes of the damages and the array of possible policies. As noted earlier, the focus is on whether the solutions are likely to be predomimantly technological or behavioral.

Air Pollution

I have suggested that the measurable costs of air pollution are probably on the order of $1,000 to $2,000 over the lifetime of a car. It

---

[12] Small (1991), pp. 43-44, 77. A similar estimated is obtained by applying the congestion cost suggested by Cameron (1991, p. 21) of $0.11/mile for Southern California to an average round-trip commute length for that region of about 20 miles.

[13] Giuliano and Small (1991), p. 169, Table 2; data are for 1980.

is no surprise, then, that people have supported technological solutions, which have proven highly effective at costs within that range or less. The passage of recent federal and state legislation in the U.S. suggests that people continue to support this approach, and are willing to extend the technological requirements from the vehicles themselves to the chemical formulation of gasoline.

But by the same reasoning, people will not support policies that they consider far more onerous than this cost. Major behavioral shifts to substantially reduce automobile use are exactly this kind of policy. The upper cost estimate just stated ($2,000 per car) amounts to approximately $0.013 per vehicle-kilometer. Cost differences far exceeding this amount have already been encountered in recent years, both among countries (especially U.S. versus Europe) and over time (especially pre-1972 versus 1980). Yet the effects of these cost differences on automobile use are very modest, especially once differences in land-use patterns are accounted for; and there appear to be no effects at all on the upward trends in automobile use. Instead, these cost differences have produced technological changes: the higher gasoline costs of the middle and late 1970s had substantial effects on vehicle size and fuel economy, but little if any on total use of personal vehicles.

Two other policies proposed for pollution control are better maintenance and inspection of motor-vehicle engines, especially the pollution-control devices; and the use of alternative fuels such as methanol. Maintenance and inspection has long been supported by professionals as a low-cost and effective policy, and seems a likely candidate for strengthening in the near future; although it requires some modest behavioral changes, it is primarily a fine-tuning of policies designed around improved technology. Alternative fuels are also more technological than behavioral, and in any event appear likely to raise costs by more than $0.013 per vehicle-kilometer, hence they are of doubtful attractiveness as a widespread strategy for reducing air pollution.

I conclude that major behavioral changes in response to automotive air pollution are unlikely. People value personal mobility too highly, and satisfactory measures are likely to be found through improved technology.

Greenhouse Effect

What sorts of reductions in automobile travel might result from reasonable policies aimed at the greenhouse effect? As noted earlier, estimates for a major reduction in CO_2 emissions suggest the incremental cost of the most expensive measure would be $0.13 per kilogram of carbon removed from the atmosphere. For the automotive sector, this is equivalent to a fuel-price increase of that amount. Based on our earlier discussion, then, we can address the question by asking how the public would respond to a fuel-tax increase of that amount.

The answer is inherent in the same natural experiments cited in the previous subsection, especially the rise in fuel prices during the 1970s. Although the total expenditure envisioned by such a drastic policy is very large, past behavior in the face of fuel-price increases of this magnitude suggests that the response is almost entirely technological; in this case, through improving fuel efficiency. The behavioral component is mainly limited to purchasing smaller vehicles.

Given the possibilities for improvning fuel efficiency at moderate cost, it seems unlikely that people will accept policies that drastically alter their mobility. This is all the more true because the direct benefits to citizens of any one nation from controlling greenhouse emissions are likely to be perceived as small. Nordhaus (1991b) estimates these benefits at just $6 billion per year for the U.S., not enough to produce political support for drastic changes; in underdeveloped nations, where the costs of climate change may be larger, resistance to any measure that would slow industrialization by restricting the use of coal and petroleum is very great.

Congestion

The technological approach to solving congestion involves better control and management of vehicles and roads, through such means as entrance-ramp metering, driver information systems, and the hoped-for payoff from developments in intelligent vehicle-highway systems (IHVS, more commonly known as "smart roads" and "smart cars"). The latter technology promises ultimately to reduce vehicle spacing and thereby increase the capacity of a given highway.

I expect that future generations will look back in disbelief at the present socially promoted, highly integrated transportation system with several million uncoordinated operators, each with little training and no ongoing supervision, operating under unknown levels of stress, alertness, and drugs, and working on headways of

1-2 seconds and lateral clearances of 1-2 meters. (In California, they do this at 100 kilometers per hour while talking on the telephone.) So IVHS will almost certainly appear in some form.

Nevertheless, IVHS technology is only peripherally related to congestion. Congestion occurs not because of poor control of vehicles, but because it is the mechanism used, by default, to ration scarce capacity. There are not enough financial resources to augment that capacity sufficiently to eliminate congestion, even if environmental considerations would permit it. Nor is there evidence so far that IVHS will prove to be a cheaper way to augment capacity. Therefore, the technological approach to congestion will probably not solve the problem, although it may have other benefits.

This leaves only two possibilities for dealing with congestion. People may simply learn to live with it, and accept it as one of life's nuisances. This is becoming easier as cars are increasingly equipped with (non-CFC-using) air conditioners, stereo sound systems, talking books, mobile telephones, fax machines, and even microwave ovens. Alternatively, people may finally accept policies that limit their freedom to travel on the most congested highways during peak hours. Such policies include parking restrictions, parking charges, carpooling incentives, and peak-hour road-use fees such as are used in Singapore and have recently been proposed in southern California, Holland, Stockholm, and Cambridge, England.

Such measures induce important behavioral shifts, forcing some people to change the time of day, mode, vehicle occupancy, or amount of their travel. Why would people tolerate policies affecting their everyday behavior in order to solve problems of congestion, but not air pollution or the greenhouse effect? There are three reasons. First, the technological alternative is not available for solving congestion, assuming car comforts alone are not enough. Second, the costs of not changing – that is, the costs of continued congestion – are very large; they are also borne directly by the traveler, so there can be no ignoring them. Third and most important, *the behavioral shifts required to alleviate congestion do not require people to give up the individualistic mobility that characterizes automobile travel.* They need only shift it out of peak periods. Hence people may respond by altering their work and travel schedules, forming carpools for peak trips only, or utilizing transit systems oriented just to peak travel. Solving congestion through behavioral shifts need not encompass their activities so broadly as would solving pollution through similar means.

CONCLUSION

The effects of transportation on air quality are potentially very serious, and justify significant mitigating or preventative measures. The same may well be true of greenhouse warming, although the scientific evidence is only in its infancy. Over the coming decades, transportation policies will be forced increasingly to take this into account.

The form which these preventative measures are likely to take is mainly technological: better pollution-control equipment, better energy efficiency, reformulated gasoline, perhaps some limited substitution of alternative fuels. Behavioral shifts are likely to be limited to better citizenship in keeping equipment working correctly. People need not and will not choose solutions that reverse the trends toward increased mobility via personal vehicles.

This contrasts with what is the only feasible solution to congestion (other than just living with it, which remains a possibility): direct incentives to reduce driving in congested areas during peak hours. These incentives, if adopted, will result in a variety of shifts which may include greater use of carpooling and public transit as well as rescheduling of activities and travel to off-peak hours. For the most part, people will nevertheless retain their demonstrated preference for private vehicles, accepting measures that alter their behavior only in very selective ways.

Those interested in strong environmental policy must, then, learn to focus their efforts within a context of affluent, highly interconnected societies in which personal control of mobility is very highly valued and will be a dominant motivation in policy formulation

REFERENCES

Abelson, Philip H. (1990), "The Asbestos Removal Fiasco", Editorial, *Science*, 247 (2 March), 1017.

Brajer, Victor, Jane V. Hall, and Robert Rowe (1991), "An Integrated Approach to Benefits Assessment: Attaining Ozone and Particulate Standards", *Contemporary Policy Issues*, 9 (1), 81-91.

Cameron, Michael (1991), *Transportation Efficiency: Tackling Southern California's Air Pollution and Congestion*. Sponsored by the Environmental Defense Fund and the Regional Institute of Southern California. Oakland, California: Environmental Defense Fund.

Crandall, Robert W., Howard K. Gruenspecht, Theodore E. Keeler, and Lester B. Lave (1986), *Regulating the Automobile*. Washington D.C.: The Brookings Institution.

Davis, Stacy C., Deborah B. Shonka, Gloria J. Anderson-Batiste, and Patricia S. Hu (1989), *Transportation Energy Data Book: Edition 10*. Oak Ridge, Tennessee: Oak Ridge National Laboratory. Distributed by U.S. Department of Commerce, National Technical Information Service, Springfield, Virginia.

Giuliano, Genevieve, and Kenneth A. Small (1991), "Subcenters in the Los Angeles Region". *Regional Science and Urban Economics*, 21 (2), 163-182.

Gordon, Deborah (1991), *Steering a New Course: Transportation, Energy, and the Environment*. Cambridge, Mass.: Union of Concerned Scientists.

Haugaard, John (1981), "Measures of Air Pollution Costs Attributable to Motor Vehicle Travel", Technical Report 134, Institute of Urban and Regional Research, University of Iowa, Iowa City (December).

Jorgenson, Dale W., and Peter J. Wilcoxen (1991), "Reducing US Carbon Dioxide Emissions: The Cost of Different Goals", Discussion Paper No. 1575, Harvard Insitute of Economic Research, Harvard University, Cambridge, Massachusetts (October):

Krupnick, Alan J., and Paul R. Portney (1991), "Controlling Urban Air Pollution: A Benefit-Cost Assessment", *Science*, 252 (26 April), 522-528.

Lave, Charles (1990), "Things Won't Get a Lot Worse: The Future of U.S. Traffic Congestion", working paper, Institute of Transportation Studies, University of California, Irvine.

Nordhaus, William D. (1991a), "The Cost of Slowing Climate Change: a Survey", *Energy Journal*, 12 (1), 37-65.

Nordhaus, William D. (1991b), "To Slow or Not To Slow: The Economics of the Greenhouse Effect". *Economic Journal*, 101 (July), 920-937.

Renner, Michael (1988), "Rethinking the Role of the Automobile", Worldwatch Paper No. 84, Worldwatch Institute, Washington, D.C.

SCAG (1988): Southern California Association of Governments,

"Congestion in the Los Angeles Region: Costs under Future Mobility Strategies". Paper prepared by the Economic Planning and Development Program. Los Angeles SCAG, March.

Small, Kenneth A. (1977), "Estimating the Air Pollution Costs of Transport Modes", *Journal of Transport Econonics and Policy,* 11 (2), 109-132.

Small, Kenneth A. (1991), *Urban Transportation Economics.* A volume in the *Fundamentals of Pure and Applied Economics* series, Regional and Urban Economics section. Chur, Switzerland: Harwood Academic Press.

U.S. Congressional Budget Office (1988), *New Directions for the Nation's Public Works.* Washington, D.C.: U.S. Government Printing Office.

U.S. Federal Highway Administration (1982), *Final Report on the Federal Highway Cost Allocation Study.* Washington, D.C.: U.S. Government Printing Office.

U.S. National Council on Public Works Improvement (1988), *Fragile Foundations: A Report on America's Public Works.* Washington, D.C.: U.S. Government Printing Office.

HUMAN ASPECTS ON COMMUNICATION AND THE INFRASTRUCTURE

Kerstin Lindahl-Kiessling
The University of Uppsala
Sweden

It seems quite clear that we are willing to overlook the comparatively high risks of travelling by road.

Around 300,000 people are killed each year in car accidents and the number seriously crippled is even larger.

Each year the active life of more than 10 million people in the world is put to an end as a result of car accidents.

In Sweden more than 20,000 people are injured on the roads every year and around 800 people are killed. These figures does not only involve car drivers, but innocent passengers, by-standers and pedestrians as well. This does not cause much commotion and very seldom any big headlines, but when three children are at risk for an unusual disease, we consider to vaccinate the whole population. Why do we set priorities like that?

Other traits of our craze for cars are more intelligible.

One very rational and understandable reason is of course the costs – I myself would almost certainly have traveled by car to this conference, instead of by 1st class railway carriage if I had to pay for my own fare. How on earth is it possible that individual transportation is cheaper than collective mass transportation on rail?

Simple. Road transportation never payed its fare share from the very beginning. If it had, many kinds of transportation would never have started and others would have been arranged in a very different way.

Who would for example get the idea to distribute fresh bread by lorries from a bakery in the very south to all over Sweden, the polar circle even, and bust all the small bakeries in the process? Or who would ever consider a bridge over Öresund, when perfectly

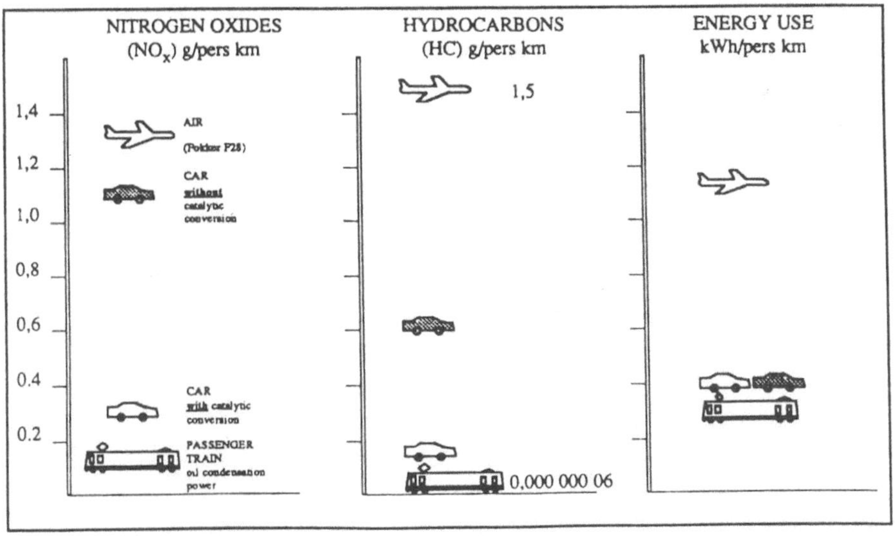

Figure 1. *Emission and energy use for different transport modes.*

acceptable and less risky alternatives are available? Who wants the bridge and who needs it?

Maybe the most important reason for such transportation anomalies is the fact that environmental costs have up till very recently been considered external features and "free commodities" in most economic calculations, which has brought us to where we are now, with road traffic as our most heavy pollutor.

As Figure 1 demonstrates the emission from old cars was of course immensely much higher per person kilometers than the train, but even the modern clean car doubles NO_X. Also for hydrocarbons the emission amount is highly dependant upon how choices between various modes of transportation and technologies are done.

Furthermore, we must realize, that the improvement of individual car performances doesn't help much due to the ever increasing number of vehicles. In spite of the newly awakened legislative and technological activity in many countries an increasing part of the air pollution comes from our cars.

These toxic substances are emitted locally where we walk, and are most concentrated right on the side walk, at a height of about one meter. The obvious question cannot be avoided – how tall are your children or your grand-children?

Figure 2. *Car trip to the mountains.*

The same increase of private car owners has of course begun in eastern Europe. I do hope they can curb this increase in proper time, earlier, much earlier than we did.

Also, the favoring of just-in-time service before transport costs alone can increase the reliance on trucks in all parts of the world, unless, as discussed by Dr. Schonberger, quality-delivery management will have the opposite effect.

If the global car explosion, had been fully met by expansion of the road network – which is fortunately enough not the case – we might have experienced an even worse environmental crisis. At present about 2 percent of the available global land is covered by asphalt or the like. Multiply that and we may begin to interfere seriously with the processes that replenish our natural water supply.

As for energy, half of the world's oil production is consumed by a fleet of 600 million road vehicles, whose growth in number has consistently outpaced that of the human population.

Which implies that even if technology makes the car motor more efficient or if people choose smaller cars, as has happened to some extent, the transport share in total oil consumption will still increase according to these calculations. Today it is close to 50 percent.

Figure 3. Traffic jam – how we use our cars most of the time.

Consequently, the transport sector is responsible for a growing share of the overall man made emissions with more or less local and regional effects, as discussed earlier, but also causes around 25 percent of the carbon dioxide emission, the potential impact which concerns the whole planet through the green house effect.

As very little efficiency improvements is occurring on a world wide basis, the motor vehicle CO_2-emission will skyrocket in the coming decades if we don't do something very drastic about it.

So much for "external features" and their so far negligible contribution to differential transportation costs. Things are maybe about to change. The programmes for automotive emission control, which began in California three decades ago and are now established in all industrialized countries, might in the long run adjust the logistics of economically motivated transportation. Statistics and several contributions to this book imply that if we create the proper incentives, there is a very large potential for improvements of goods provision logistics.

But the private car is a different matter. Its strongest reason for existing is certainly not the economical aspect but the unsurpassed mobility it gives. It's a well known fact.

I don't need to spend time on describing the prolog to Figure 2 how I put the skis on the rack before I put my luggage, my kids and my dogs into my car and start out for the mountains.

This, however, is not how we use our cars most of the time – but the queuing-up as shown in Figure 3.

In fact, we don't need these strong, heavy, fast cars. A cheap little car would take you there just as slowly. The rational thing would be a small electric car for the city. When I want to go to my summer house or some place else more distant, I would load it as usual at home and drive to the railway station. At the station there would be a railway carriage with both front and end openings, similar to how most ferries are constructed. I would park my car in this carriage and myself in the train restaurant. Three hours or so later the train stops in Järvsö 300 kilometers north of Stockholm, where I after a tasty meal and a nap or a social chat would get into my car, dismount and drive the few remaining kilometers. I look forward to that kind of more humane transportation possibility.

Unfortunately, the small rational car has so far not been able to vanquish the market. Most certainly because costs and mobility alone does not explain our infatuation and love for cars and especially large and fast ones. Cars have always been connected with a gilt-edged glory, which cannot be fully understood intellectually, as visualized in this picture.

Table 1. *How get to work in different cities*

City	Land Use Intensity (pop+jobs/ha)	Private car	Public transport	Walking and cycling
		(percent of workers using)		
Phoenix	13	93	3	3
Perth	15	84	12	4
Washington	21	81	14	5
Sydney	25	65	30	5
Toronto	59	63	31	6
Hamburg	66	44	41	15
Amsterdam	74	58	14	28
Stockholm	85	34	46	20
Munich	91	38	42	20
Vienna	111	40	45	15
Tokyo	171	16	59	25
Hong Kong	403	3	62	35

Maybe density itself will necessitate different, more humane modes of transportation. Car density is not identical with prosperity, as Table 1 shows. Perhaps the solution to the "sausage" or "banana" problem mentioned by other contributors to this book has already been found in Hong Kong, where 35 percent of the people cycle to work compared to 3 percent in Phoenix, or in China where there are 250 bicycles for every car compared to 0.7 in the USA.

However, "if we don't understand what constitutes this deep desire for owning a car we will not be able to change the pattern" (Emin Tengström). So we better find out soon. This is obviously a male dream, however. Therefore, in my opinion it is the dominating male influence in this sector which is the main obstacle for a more rational development, but I have no proof for that. If I am right, there may be a drastic change very soon because women are most likely the largest group of car owners-to-be in the near future. So the car producers should watch out! Be prepared for a rush when someting really different enters the market.

It is not all darkness though. There are new options. New technological options are around the corner, so we have heard. Already, today about 50 million cars are not using pure petrol any longer, and thus do not pollute the air as much and don't contribute to the world's energy and carbon dioxide crisis.

Figure 4. *Infatuation and love for cars.*

I am sure that various transport management measures and institutional arrangements can contribute immensely towards a sustainable development. Means to that effect are in your hands, dear reader, top administrators and highly ranked officials or entrepreneurs with great power. Or when we united as millions of customers are asking for something, it is bound to happen.

You must never forget that the transport sector is the major contributor to the environmental threat and thus it has to pay for and take the responsibility for changes, which means that we must set considerably more ambitious goals than those currently existing.

I will confine myself to goals most closely connected to the Swedish Road Administration and its counterparts in other countries.

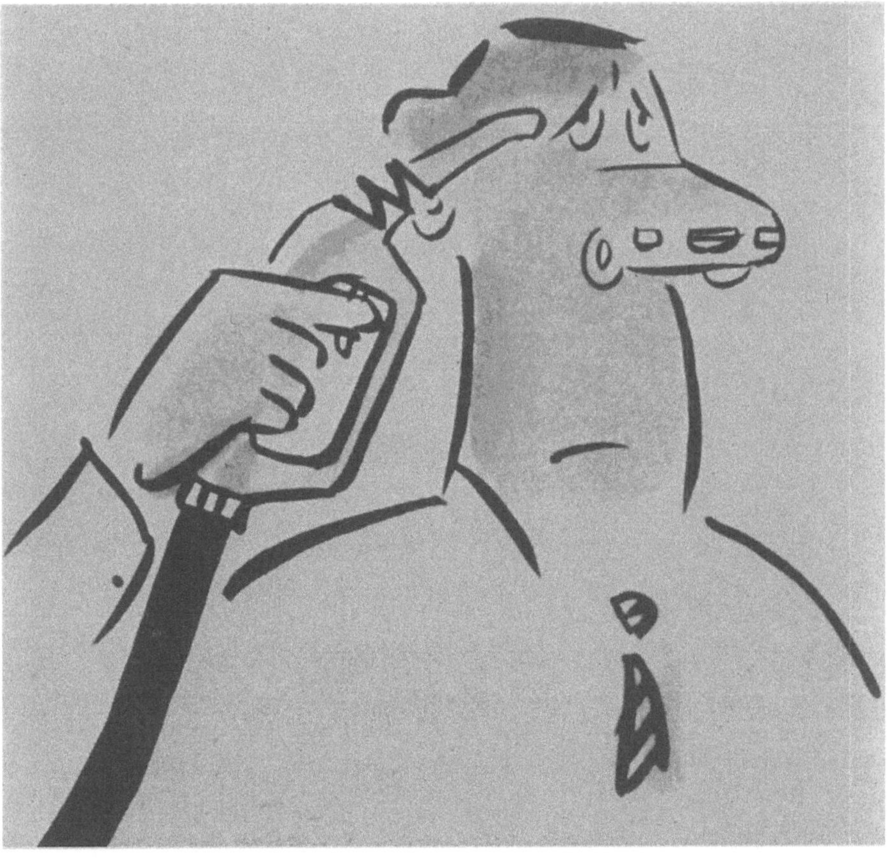

Figure 5. The Artist's eye upon human being and the car.

Figure 6. *Technology and inventiveness can do wonders.*

It is certainly a global need for heavy investments in transport infrastructure in most countries throughout the world. There is also an urgent need for a change of priority in favor of more environmentally and friendly modes of transportation. Here physical planning and traffic management are important instruments for promoting a development with sound and clean transport infrastructure and lower transport demand.

I think that if road authorities in all countries don't want to be left behind, they must no longer take their point of departure in projections of the expected increase in the number of cars.

If we take such exponential forecasts as the basis for planning, we will contribute to self-fulfillment and we may soon find that the car explosion has become as severe a problem as population explosion has ever been.

We have to start in the opposite end – that is our starting point must be consideration for natural resources and our environment. If not, we may shorten our prospects considerably below seen through the artist's eye.

Of course, I don't want to get rid of the beloved car or go back to the stone age. At least not seriously and 365 days a year. Certainly, this will not happen. There are too many options, that can be developed by clever inventors, researchers, businessmen and administrators – only a new attitude has to be firmly rooted in all planning and decision-making processes.

The obvious rationale why we have to consider the natural resources as our basis instead of the projected number of cars, when planning how many and which new roads to build, is the fact that technology and inventiveness can do wonders, take us to the moon to explore universe.

We cannot learn how to create a new globe. This is the earth we have and we have to take good care of her. To accomplish that, we need a revolution, no less, in our attitude to mobility and transportation.

A revolution, using all the means there are, behavioral, technological, even legislative, to finally achive what should have been our goal right from the start: the environmentally sound solution to every transportation problem shall always be made the most economical and the most attractive.

REFERENCES

International and Swedish statistics and official reports.

Brown, Lester et. al.. *State of the world*. 1991

Ehrlich, Anne H. and Ehrlich, Paul R.. *Earth*. 1987

Månsson, Tommy,. Transport and Environment, A contribution to UNEP's State of the Environment Report 1972-1992. (To be published)

Tengström, Emin. *Bilismen — i kris?* 1991. (Swedish)

Älskade hatade bil, Forskningsnämnden 1990. (Swedish)

TECHNOLOGY FOR IMPROVED TRANSPORTATION AND ENVIRONMENT

Hans G. Forsberg
President of the Royal Swedish Academy of Engineering Sciences
Sweden

Let me start by emphasizing something which has been said a great many times: the earth's resources are very unequally distributed in relation to its population. This is true on a national, regional and global level. In the highly specialized, industrialized societies of the Western world the extensive transportation of people and goods is absolutely vital. The more efficiently such transportation can be organized, the more assets are created or liberated to meet different human needs, be they material or cultural.

There is no doubt among economic historians that the most important technical achievements of the last century were those which made trade and transport easier and more efficient. This in turn contributed to an economic development without parallel in the history of the industrialized countries. The steam turbine, the internal combustion engine and electrical machinery made it possible to replace people and beasts of burden. The increase in productivity was enormous. Up until the turn of the century sailing vessels could hold their own, but since then competition from steam and motor ships has completely overtaken them.

There are any number of anecdotes about the sceptical attitude towards road and rail vehicles when they were first to be introduced. Laws were passed decreeing that people with red flags should precede the vehicles in order to warn other travellers, and it was claimed that people would go mad if they were to see or travel in trains and other self-propelled vehicles rushing through the countryside at high speeds. Well, safety still leaves a lot to be desired and travelling undoubtedly has a profound effect on our mental outlook.

In recent years justified criticism has been directed at the various types of health and environmental problems which traffic, as well as other industrial activities, has indisputedly caused. Some of the increased resources created thanks to transportation have been lost, at least in economic terms, owing to environmental deterioration.

For as much as a quarter of a century planners and legislators have been surrounding transport and traffic with restrictions out of regard for the environment. Some of these were perfectly justified, others no more than empty gestures.

But where do the problems arise?

All combustion of fuels containing carbon gives rise to carbon dioxide, which together with other gases can lead to the so-called greenhouse effect. As we have already heard, the carbon dioxide content in the atmosphere is increasing by 0.4 percent each year. Traffic is calculated to account for about half of this increase.

When combustion is incomplete, then carbon monoxide, soot and a number of chemical compounds are formed. Many of them have a pronounced effect on our health when we inhale them, but they also give rise to smog, which is a regional environmental problem. Nitrogen oxides belong in the same category – they are always formed to some extent when combustion takes place using air. Fuels containing sulphur give rise to sulphur dioxide which, together with nitrogen oxides, is responsible for local and regional acidification causing direct damage to plant life and buildings.

For many years now consumers, legislators and manufacturers have in fact all shared the view that we must work to reduce emissions from traffic. Two trends can be discerned. One is concerned with efforts to improve transport efficiency; that is – to reduce fuel consumption and thus the emission of environmentally damaging substances. The tremendous price fluctuations of crude oil during the 1970s, the repeated warnings about the lack of suitable oil fractions for transportation purposes and the discovery by politicians that you can tax gasoline have all been powerful incentives to improve efficiency. Streamlined bodywork was one of the first methods used to reduce energy consumption, even if a car's design was mainly determined by aesthetic ideals. Naturally, the price of gasoline can be almost sacred in some cultures. A tax increase on gas in America to finance the war in Kuwait was unthinkable for President Bush, even though vital oil interests were at stake.

The second trend is concerned with the combustion process itself, or with the chemical modification of the products resulting from combustion so that emissions are rendered harmless. Major advances have been made in this field in the last 20 years. Thanks to the electronic control of combustion and catalytic cleaning, it has been possible to render emissions from Otto engines driven at a constant load - for example, when travelling at a constant speed on motorways - practically free from environmentally damaging gases, with the exception of carbon dioxide. At the same time lead tetraethyl has been replaced by other octane-increasing substances. 90-95 percent of emissions can be dealt with in this way. What remains? Cold starts and rapid accelerations.

Many manufacturers and users of the Otto engine are engaged in continuing development with a view to a further reduction of emissions. In our Swedish climate there are many who for reasons of comfort choose to keep both the engine and the interior of their car warm with an electric heater. If you do this you can reduce the emissions resulting from a cold start as long as the catalyst is also preheated. The prevailing system with an oxygen sensor and three-way catalyst can be further improved. The combustion process itself can be made more efficient and electronics can adjust the ignition process and report on how well the catalyst is functioning. Those in favour of automatic gear boxes point out that these can admittedly mean reduced engine efficiency and subsequently increased fuel consumption, but that this is more than compensated for by the fact that acceleration is smoother. This means that the electronic systems and the catalyst can cope more effectively. Cars straight off the production line from the majority of car factories now surpass current national regulations by a good margin. If car manufacturers are forced to meet even tougher regulations then it will presumably be necessary to change the composition of the fuel. I will get back to this later.

The Diesel engine, which has a higher combustion efficiency than the Otto engine and which is able to use heavier, cheaper oil fractions, produces less emissions of hydrocarbons and carbon monoxide. On the other hand there are problems with unburnt particles and nitrogen oxides. Solving one of these problems usually means that we encounter increased difficulties with the other.

The direct injection supercharged diesel engine is the dominant method of propulsion for heavy duty vehicles (trucks and buses). For a number of years now development work has been conducted to obtain higher efficiency and lower fuel consumption, which in themselves lead to lower emissions. But it is really in the last ten years or so that technical advances have succeeded in reducing emissions with more radical methods.

The catalyst techniques with the oxygen sensor which have been successfully applied to the purification of exhaust emissions from the Otto engine cannot be used on exhaust emissions from diesel engines because of the way in which the diesel engine works (with excess air). Carbon and other deposits on the catalyst are also obstacles.

All heavy-vehicle manufacturers with any self respect are also conducting development projects to solve the problems of nitrogen oxides.

One example is the introduction of a cooling system for the supercharged air which gives lower combustion temperatures and thus less nitrogen oxides: you can have delayed fuel injection too, but this means increased fuel consumption. Improved techniques for fuel injection and electronic control of the injection processes can give some positive changes here. But even so, we are still only talking about reducing nitrogen oxide emissions to about one half of today's levels.

However, it is possible to achieve a greater reduction using different versions of the compound engine.

In order to deal with unburnt carbon it would seem that manufacturers must mount special filters in the exhaust system. The particles can then be burned off by a special cleaning process. However, the filter becomes blocked quite easily and has to be regenerated rather frequently. Volvo has introduced its City Filter which needs to be cleaned after 300-500 kilometres; this is a simple matter for a bus on urban routes – you just do it at night.

Certain changes in the composition of fuel can also lead to a reduction in emissions of particles. Simply switching to sulphur- and aromatic-free diesel oil means that significantly less carbon particles are emitted.

Over the years many interesting types of engines have seen the light of day and tried to compete with the two established concepts: the Wankel Engine, the Stirling Engine, gas turbines and so on. But I think I am right in saying that, even if they seen to offer low emissions in certain circumstances, they present many other problems which mean that we cannot expect to see them introduced commercially within the forseeable future. The same is true of the interesting idea of re-introducing steam engines. I remember the so called steamcar from the beginning of this century.

Liquid hydrocarbons have several practical advantages. Even if gasoline is highly inflammable we have learned how to handle it. There are extremely few accidents these days in refineries or with tankers at gasoline stations. One way of reducing emissions has been in use for some time – lighter hydrocarbons such as butane and propane or a mixture of these two gases have been popular fuels for trucks and taxis in some cities. LPG (liquefied petroleum gas) is on sale for use in ordinary cars in many gasoline stations in different parts of the world.

Experiments are now underway – mainly in the US and New Zealand – with natural gas, that is methane. Today, in a number of cities, in the United States, 1,500 heavy vehicles – mainly garbage trucks and school buses – run on natural gas. Here we are talking about conventional vehicles which have been converted. The disadvantage with natural gas is that in most countries there are only a few places where one can build the compressor stations necessary to "fuel" the vehicles. Thus, vehicles that run on natural gas will almost certainly remain a curiosity in most countries even if the emissions from such vehicles are guaranteed free from heavy hydrocarbons and nasty carcinogenic substances produced by combustion and particles. However, the formation of nitrogen oxides is at least as high as in conventional combustion processes.

The use of alcohols instead of or as a supplement to gasoline is not a new phenomenon. We used it in Sweden during the blockade in the Second World War and in Brazil they have been using ethanol as a supplement for decades. The Brazilians also have the well known Alcohol Program with straight (neat) ethanol; this is a way for them to utilize their sugar surplus. However, this program, currently faces difficulties, mainly because of cheap oil. The two alcohols which are of interest here are ethanol and methanol, both of which can be made from crude oil but also chemically or by microbiological processes from vegetable or cellulose materials.

Alcohols generally have lower combustion values per unit than gasoline because of the oxygen atom in the molecule. On the other hand, the alcohol supplement raises the octane number, which means that compression can increase and with this combustion efficiency. There is great interest in working with these fuels in different countries. There is a considerable agricultural surplus and many people are convinced that the price of fossil fuels will rocket in the next few decades. In a global perspective there is also concern about the carbon dioxide which is released during the burning of agricultural and forest products. But these would have ultimately been released into the Earth's natural cycle when the products were eaten by humans and animals or when they just rot down. However, I think I can safely say that it will take a decade or so before we see vehicles running completely or partially on methanol or ethanol on any large scale. There is still little or no market pull. It will also be a while before fuels containing alcohol supplements are generally available other than the mixtures containing a small percentage of methanol-based ethers which are already on the market.

At regular intervals proposals for radical changes in the transport system are put forward: a switch from liquid fuels to electricity or hydrogen. How are such ideas making out?

Electrically powered vehicles are hardly a novelty. The electrification of our railways and tramways started at the beginning of this century. The underground systems which have been built for local transport would be quite unthinkable today without electricity. Trolley buses are still operating in many parts of the world, even if they are less flexible in the traffic system than ordinary buses. In mines and factories, electrically powered trucks have taken over – these are either run via cables, current collectors, conductors or on batteries. In such situations it is, as a rule, possible to recharge or change the batteries in a simple and efficient way.

Many people have tried to come up with an electrically powered car for use in urban traffic. In Europe, Japan and the United States a number of test programmes are underway. Electrically propelled cars and buses will be tested for a period of 5-10 years. The problem is still the range. How far can I drive before the battery needs recharging? A practical range with the batteries available today is between 100 to 200 kilometres. Electric propulsion is attractive from an environmental point of view: vehicles are both silent and free from exhaust pollutants. Naturally there is then the question of the environmental problems which arise at the power stations that generate the electricity used to charge the batteries. I shall come back to that.

I must confess that I personally am fascinated by the hybrid cars which run on different combinations of internal combustion and electric power. For long distance travel for example you can use an Otto engine partly for propulsion and partly to charge the batteries. The batteries can then be used to propel the car when travelling in heavy urban traffic where there are local restrictions on emissions. But experiments are also being made with continuous electric power, in which a combustion engine with a constant engine speed is used only for charging the batteries which then propel the car with an electric motor. Such a system offers good acceleration without increasing emissions. The practicality of this concept is demonstrated in many other applications: diesel electrical machinery in ships and Stirling-engine-operated submarines in the Swedish navy. It is also possible to utilize brake power to charge the batteries. Such a technique corresponds to the various experiments which have been carried out to improve the operating economy for buses in urban traffic using flywheels or compressed air tanks.

In many scenarios for a time in the future when crude oil is in short supply or when the greenhouse effect forces us to make international agreements to radically reduce carbon dioxide emissions, propulsion based on hydrogen gas appears to be an environmentally attractive alternative. Modifying today's international combustion engines to suit an explosive gas mixture does not appear to present unsurmountable difficulties. Nor should it be too difficult to create an exchange or refill system for hydrogen cylinders. A system where the hydrogen is combusted using oxygen as opposed to air has a special appeal. If this can be done then it is also a perfect way of avoiding the formation of nitrogen oxides.

The only reasonable way we know of today for producing hydrogen gas is by electrolysis of water. This requires large amounts of electricity. Now we come back to the question of how electricity can be produced without an undue negative impact on the environment. Hydropower is extremely cheap and we still have massive, unexploited hydropower resources. But in many parts of the world powerful environmental interests are opposed to further construction projects. Nuclear power is appealing and fairly cheap but that too has its opponents. Using fossil fuels or wood fuel to fire large installations for the production of hydrogen is not particularly attractive. The conversion efficiency in thermal power plants is not more than 40 percent and it may be far better from a systems point of view to burn the fuels directly in an engine. Fusion is still a possibility; but the prospect is still as distant as we thought it to be when work on fusion started after the detonation of the first hydrogen bomb. That is to say, at least 50 years more to go.

Recently some ideas have been put forward to produce hydrogen from biological material. This is exciting, but again we will need decades of research and development before we can know the economic facts of such a process.

I have not mentioned railways as alternatives to increased road transport. The reason is simple. All the technology is already here and there is much space available before the tracks of Europe are filled. To increase the amount of goods transported by rail is not a technical issue but a management problem – to build and split up a train of 50-100 units. Just-in-time delivery which industry requests more and more – even over long distances – makes necessary other business units than the present European Railway Administrations – and this may take a while.

The enthusiasts always say that the break through is just around the corner for electric cars, for non-hydrocarbon fuels, for the large-scale use of hydrogen. So they were 15 years ago after the first oil price shock – and so they may remain for a long time.

I am far from being a pessimist when it comes to technical development. So many surprising things happen all the time, things which turn ingrained ideas upside down. But I think that when its comes to developments in transport for the future it is the many small improvements which will prove decisive. There is still plenty which can be done to improve the current systems. There are still a lot of engines just ticking over on our roads – diversions and traffic jams lead to unnecessary energy consumption which in turn places an unnecessary burden on our natural environment.

Modern information technology can by applied to traffic and environmental problems in many contexts. It is important in controlling the efficiency of combustion and in the cleansing of exhaust gas already referred to. There are currently several projects in Europe and the United States in which the traffic situation and the best alternative for each individual vehicle can be reported to the driver. Tomorrow we will be able to see that the technology being developed for air traffic control will also be used to benefit at least the professional truck drivers.

Economic conditions change too. Investments in docks and railways are made in perspectives extending over a century or more; roads and bridges are built to last for several decades, while vehicles and vessels have an economic life of between 10 and 20 years. Here we can see that wear and tear with the ensuing maintenance costs, safety checks on aircraft, ships and vehicles also set physical limits.

One environmental aspect which should not be forgotten here is that re-utilization of the materials used in vehicles and vessels is now an extensive industry and that it too has its own particular environmental problems. It is a great challenge for the designers of the future to construct our various conveyances so that they are light and easy to dismantle so that the material can be reused as far as possible, for example, after it has been remelted.

So far I have not mentioned noise from traffic or those measures which can be taken to reduce it, but there are several ways in which these problems can be solved. Today, it is not the engines which are the nuisance, it is the noise from the tyres which is most troubling. Here we see a challenge for tyre manufacturers and road engineers, which calls for closer collaboration.

Safety in and outside the vehicle will also be improved. But here it is not just a matter of technical solutions. The psychological aspects of driving have to be more thoroughly understood.

Health and environmental problems in the larger urban areas will require new system approaches. Our old city centres were quite simply not built to meet with the present day traffic situation. Neither hybrid nor electrically powered cars will help to relieve the congestion we experience today. In densely populated areas public transport systems must take first place – but outside these areas the individual's or the company's special values and choices must be respected. Environmental and traffic policies are integral parts of an overall discussion about the problems of modern society. The emissions from traffic must be weighed against the problems which arise from other human activities.

In a report from the Academy of Engineering Sciences which has just been published, we say – and I quote: "The opportunities for stable, long-term technical development are to be found in active interaction between international and national market conditions and systems of rules. The rules of society for safety and environmental requirements must interact with the technological knowledge base and build on deeper and wider awareness of technology and environment among the players on the marketplace. The long-term societal responsibilites and opportunities are related above all to stepping up of the R&D efforts. This will also add strength to the long-term incentives for industrial research and development."

REFERENCES

Swedish Academy of Engineering Sciences. 1991. *Technology for Transport and Environment*. IVA-M269. Stockholm.

MIX
Papier aus verantwortungsvollen Quellen
Paper from responsible sources
FSC® C105338

If you have any concerns about our products,
you can contact us on
ProductSafety@springernature.com

In case Publisher is established outside the EU,
the EU authorized representative is:
Springer Nature Customer Service Center GmbH
Europaplatz 3, 69115 Heidelberg, Germany

Printed by Libri Plureos GmbH
in Hamburg, Germany